Made in the United States
Text printed on 100%
recycled paper

Houghton
Mifflin
Harcourt

Printed in the U.S.A.

ISBN 978-0-544-34214-9

2 3 4 5 6 7 8 9 10 0928 22 21 20 19 18 17 16 15 14

4500476012   ^ B C D E F G

Dear Students and Families,

Welcome to **Go Math!**, Grade 3! In this exciting mathematics program, there are hands-on activities to do and real-world problems to solve. Best of all, you will write your ideas and answers right in your book. In **Go Math!**, writing and drawing on the pages helps you think deeply about what you are learning, and you will really understand math!

By the way, all of the pages in your **Go Math!** book are made using recycled paper. We wanted you to know that you can Go Green with **Go Math!**

Sincerely,

The Authors

Made in the United States
Text printed on 100% recycled paper

GO MATH!

Authors

Juli K. Dixon, Ph.D.
Professor, Mathematics Education
University of Central Florida
Orlando, Florida

Edward B. Burger, Ph.D.
President, Southwestern University
Georgetown, Texas

Steven J. Leinwand
Principal Research Analyst
American Institutes for
   Research (AIR)
Washington, D.C.

Contributor

Rena Petrello
Professor, Mathematics
Moorpark College
Moorpark, California

Matthew R. Larson, Ph.D.
K-12 Curriculum Specialist for
   Mathematics
Lincoln Public Schools
Lincoln, Nebraska

Martha E. Sandoval-Martinez
Math Instructor
El Camino College
Torrance, California

English Language Learners Consultant

Elizabeth Jiménez
CEO, GEMAS Consulting
Professional Expert on English
   Learner Education
Bilingual Education and
   Dual Language
Pomona, California

Fractions

 Common Core **Critical Area** Developing understanding of fractions, especially unit fractions (fractions with numerator 1)

 Critical Area

8 Understand Fractions   441

COMMON CORE STATE STANDARDS
3.OA Number & Operations
Cluster A Develop understanding of fractions as numbers.
3.NF.A.1, 3.NF.A.2a, 3.NF.A.2b, 3.NF.A.3c

GO DIGITAL

Go online! Your math lessons are interactive. Use *iTools*, Animated Math Models, the Multimedia eGlossary, and more.

Essential Question

? What are equal parts of a whole?

Start

Chapter 8 Overview

In this chapter, you will explore and discover answers to the following **Essential Questions**:

- How can you use fractions to describe how much or how many?
- Why do you need to have equal parts for fractions?
- How can you solve problems that involve fractions?

Personal Math Trainer
Online Assessment and Intervention

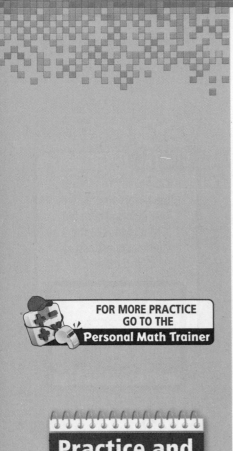

FOR MORE PRACTICE
GO TO THE
Personal Math Trainer

**Practice and
Homework**

Lesson Check and
Spiral Review in
every lesson

Critical Area | Fractions

Common Core **CRITICAL AREA** Developing understanding of fractions, especially unit fractions (fractions with numerator 1)

The Missouri quarter shows explorers Lewis and Clark traveling down the Missouri River. The Gateway Arch is in the background.

Real World Project

Coins in the U.S.

Many years ago, a coin called a *piece of eight* was sometimes cut into 8 equal parts. Each part was equal to one eighth ($\frac{1}{8}$) of the whole. Now, U.S. coin values are based on the dollar. Four quarters are equal in value to 1 dollar. So, 1 quarter is equal to one fourth ($\frac{1}{4}$) of a dollar.

Get Started

WRITE ▸ Math

Work with a partner. In which year were the Missouri state quarters minted? Use the Important Facts to help you. Then write fractions to answer these questions:

1. 2 quarters are equal to what part of a dollar?

2. 1 nickel is equal to what part of a dime?

3. 2 nickels are equal to what part of a dime?

Important Facts

- The U.S. government minted state quarters every year from 1999 to 2008 in the order that the states became part of the United States.
- 1999—Delaware, Pennsylvania, New Jersey, Georgia, Connecticut
- 2000—Massachusetts, Maryland, South Carolina, New Hampshire, Virginia
- 2001—New York, North Carolina, Rhode Island, Vermont, Kentucky
- 2002—Tennessee, Ohio, Louisiana, Indiana, Mississippi
- 2003—Illinois, Alabama, Maine, Missouri, Arkansas
- 2004—Michigan, Florida, Texas, Iowa, Wisconsin
- 2005—California, Minnesota, Oregon, Kansas, West Virginia
- 2006—Nevada, Nebraska, Colorado, North Dakota, South Dakota
- 2007—Montana, Washington, Idaho, Wyoming, Utah
- 2008—Oklahoma, New Mexico, Arizona, Alaska, Hawaii

_____

_____

_____

_____

Completed by _____

Understand Fractions

Show What You Know

Personal Math Trainer
Online Assessment and Intervention

Check your understanding of important skills.

Name _____

▶ **Equal Parts** Circle the shape that has equal parts.

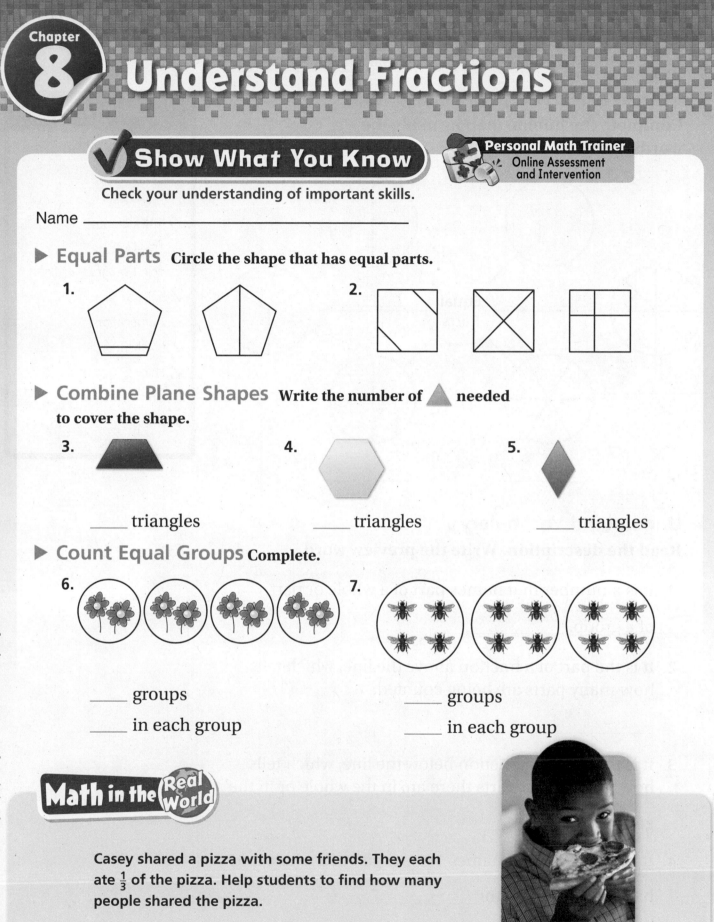

1.

2.

▶ **Combine Plane Shapes** Write the number of ▲ needed to cover the shape.

3. ____ triangles

4. ____ triangles

5. ____ triangles

▶ **Count Equal Groups** Complete.

6. ____ groups
____ in each group

7. ____ groups
____ in each group

Math in the Real World

Casey shared a pizza with some friends. They each ate $\frac{1}{3}$ of the pizza. Help students to find how many people shared the pizza.

Vocabulary Builder

▶ **Visualize It** · · · · · · · · · · · · · · · · · ·

Complete the bubble map by using the words with a ✓.

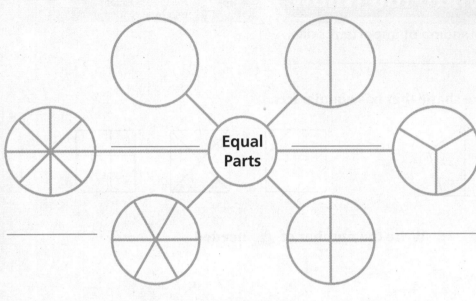

Equal Parts

© Houghton Mifflin Harcourt Publishing Company

Preview Words

denominator
✓ eighths
equal parts
✓ fourths
fraction
fraction greater than 1
✓ halves
numerator
✓ sixths
✓ thirds
unit fraction
✓ whole

▶ **Understand Vocabulary** · · · · · · · · · · · · · · · · · ·

Read the description. Write the preview word.

1. It is a number that names part of a whole or part of a group. _____

2. It is the part of a fraction above the line, which tells how many parts are being counted.

   _____

3. It is the part of a fraction below the line, which tells how many equal parts there are in the whole or in the group. _____

4. It is a number that names 1 equal part of a whole and has 1 as its numerator. _____

- **Interactive Student Edition**
- **Multimedia eGlossary**

Chapter 8 Vocabulary

denominator

denominador

11

Eighths

octavos

17

Equal Parts

partes iguales

21

Fourths

cuartos

26

fraction

fracción

27

Fraction Greater than 1

fraccíon mayor que 1

28

Halves

mitades

32

numerator

numerador

53

These are eighths

The part of a fraction below the line, which tells how many equal parts there are in the whole or in the group

Example: $\frac{1}{5}$ ← denominator

These are fourths

Parts that are exactly the same size

6 equal parts sixths

A number which has a numerator that is greater than its denominator

Examples:

$\frac{6}{3}$ $\frac{2}{1}$

A number that names part of a whole or part of a group

Examples:

$\frac{1}{3}$

The part of a fraction above the line, which tells how many parts are being counted

Example: $\frac{1}{5}$ ← numerator

These are halves

Sixths

sextos

74

Thirds

tercios

77

unit fraction

fraccíon unitaria

79

Whole

entero

84

These are thirds

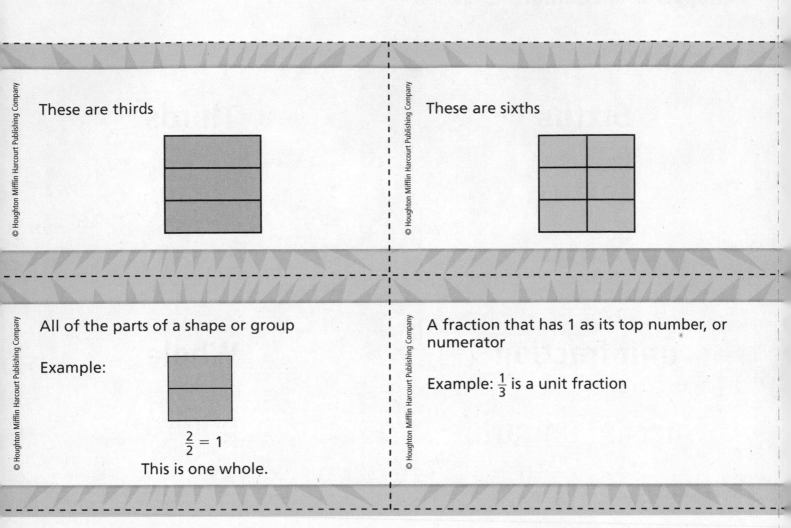

These are sixths

All of the parts of a shape or group

Example:

$\frac{2}{2} = 1$

This is one whole.

A fraction that has 1 as its top number, or numerator

Example: $\frac{1}{3}$ is a unit fraction

Going to the Mint

For 2 to 4 players

Materials

- 3 red connecting cubes
- 3 blue connecting cubes
- 3 green connecting cubes
- 3 yellow connecting cubes
- 1 number cube

How to Play

1. Put your 3 connecting cubes in the START circle of the same color.

2. To get a cube out of START, you must roll a 6.
   - If you roll a 6, move 1 of your cubes to the same colored circle on the path.
   - If you do not roll a 6, wait until your next turn.

3. Once you have a cube on the path, toss the number cube to take a turn. Move the connecting cubes that many tan spaces. You must get all 3 of your cubes on the path.

4. If you land on a space with a question, answer it. If you are correct, move ahead 1 space.

5. To reach FINISH, move your connecting cubes up the path that is the same color as your cubes. The first player to get all three cubes on FINISH wins.

Game

START

START

FINISH

What is the meaning of a **whole?**

How many sixths does three thirds equal?

Why is $\frac{1}{4}$ a unit fraction?

What is the name of the part of a fraction above the line?

How many fourths are in a whole?

What is a fraction?

Why is $\frac{4}{3}$ a fraction greater than 1?

Which is a whole: $\frac{3}{8}$ or $\frac{8}{8}$?

442B

START

What are sixths?

What kind of number has a numerator greater than its denominator?

FINISH

What kind of fraction is $\frac{1}{3}$?

If a whole has two equal parts, what are the equal parts called?

How many equal parts called thirds are in a whole?

What are equal parts?

If there are 8 equal parts in a whole, what are the equal parts called?

What is the meaning of *denominator*?

START

The Write Way

Reflect

Choose one idea. Write about it.

- Draw and explain the ideas of *equal* and *unequal parts*. Use a separate piece of paper for your drawing.
- Tell the most important idea to understand about fractions.
- Define *numerator and denominator* so that a younger child would understand.

Name _____

Equal Parts of a Whole

Essential Question What are equal parts of a whole?

Common Core Number and Operations—Fractions—3.NF.A.1 *Also 3.G.A.2*
MATHEMATICAL PRACTICES
MP2, MP4, MP5

🔑 Unlock the Problem Real World

Lauren shares a sandwich with her brother. They each get an equal part. How many equal parts are there?

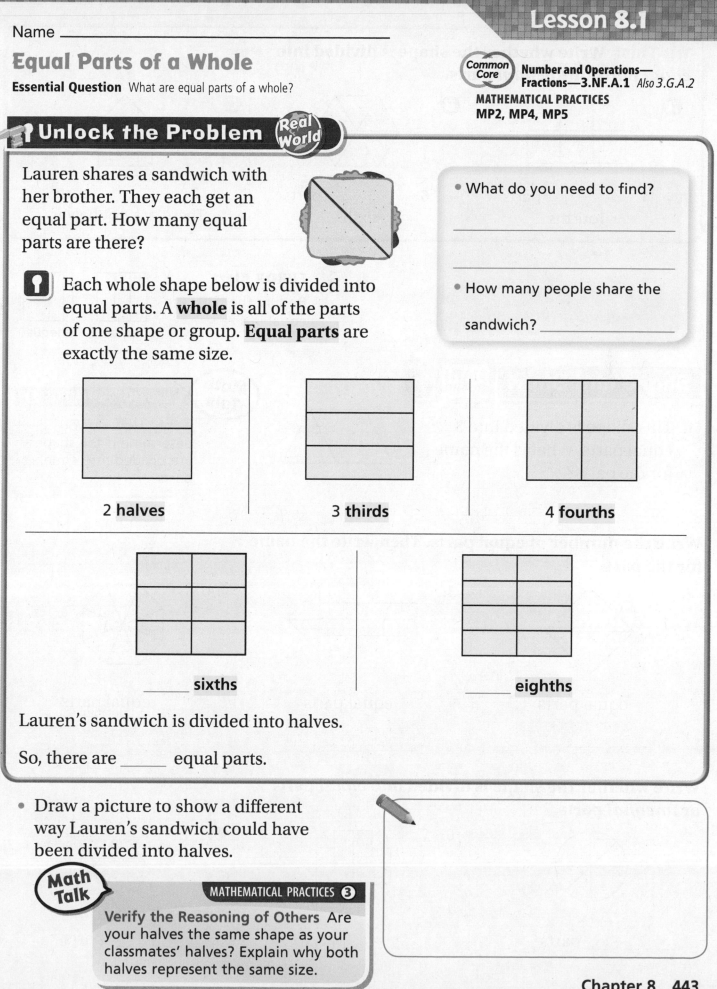

• What do you need to find?

_____

_____

• How many people share the sandwich? _____

🔓 Each whole shape below is divided into equal parts. A **whole** is all of the parts of one shape or group. **Equal parts** are exactly the same size.

2 **halves**   3 **thirds**   4 **fourths**

_____ **sixths**   _____ **eighths**

Lauren's sandwich is divided into halves.

So, there are _____ equal parts.

• Draw a picture to show a different way Lauren's sandwich could have been divided into halves.

Math Talk

MATHEMATICAL PRACTICES ❸

Verify the Reasoning of Others Are your halves the same shape as your classmates' halves? Explain why both halves represent the same size.

Try This! Write whether the shape is divided into *equal* parts or *unequal* parts.

A

4 _____ parts
fourths

B

6 _____ parts
sixths

C

2 _____ parts
These are not halves.

Share and Show | MATH BOARD

Math Talk   **MATHEMATICAL PRACTICES** ❸

Apply How do you determine if the shapes are divided into equal parts?

1. This shape is divided into 3 equal parts. What is the name for the parts?

_____

Write the number of equal parts. Then write the name for the parts.

2.

_____ equal parts

_____

3.

_____ equal parts

_____

☑4.

_____ equal parts

_____

Write whether the shape is divided into *equal* parts or *unequal* parts.

5.

_____ parts

6.

_____ parts

☑7.

_____ parts

Name _____

Write the number of equal parts. Then write the name for the parts.

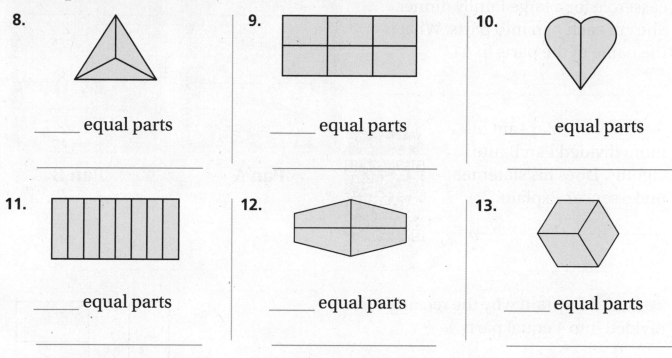

8.

_____ equal parts

_____

9.

_____ equal parts

_____

10.

_____ equal parts

_____

11.

_____ equal parts

_____

12.

_____ equal parts

_____

13.

_____ equal parts

_____

Write whether the shape is divided into *equal* parts or *unequal* parts.

14.

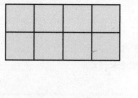

_____ parts

15.

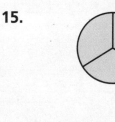

_____ parts

16.

_____ parts

17. Draw lines to divide the circle into 8 eighths.

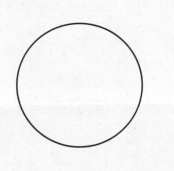

18. **GO DEEPER** Thomas wants to divide a square piece of paper into 4 equal parts. Draw two different quick pictures to show what his paper could look like.

Problem Solving • Applications

Use the pictures for 19–20.

19. Mrs. Rivera made 2 pans of corn casserole for a large family dinner. She cut each pan into parts. What is the name of the parts in A?

_____

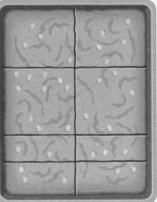

20. **THINK SMARTER** Alex said his mom divided Pan B into eighths. Does his statement make sense? Explain.

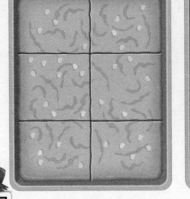

_____

Pan A   **Pan B**

21. **MATHEMATICAL PRACTICE 6** **Explain** why the rectangle is divided into 4 equal parts.

_____

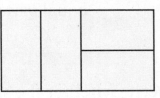

22. **GO DEEPER** Shakira cut a triangle out of paper. She wants to divide the triangle into 2 equal parts. Draw a quick picture to show what her triangle could look like.

23. **THINK SMARTER** Parker divides a fruit bar into 3 equal parts. Circle the word that makes the sentence true.

The fruit bar is divided into | thirds / halves / fourths |.

Equal Parts of a Whole

Write the number of equal parts.
Then write the name for the parts.

Common Core

COMMON CORE STANDARD—3.NF.A.1
Develop understanding of fractions as numbers.

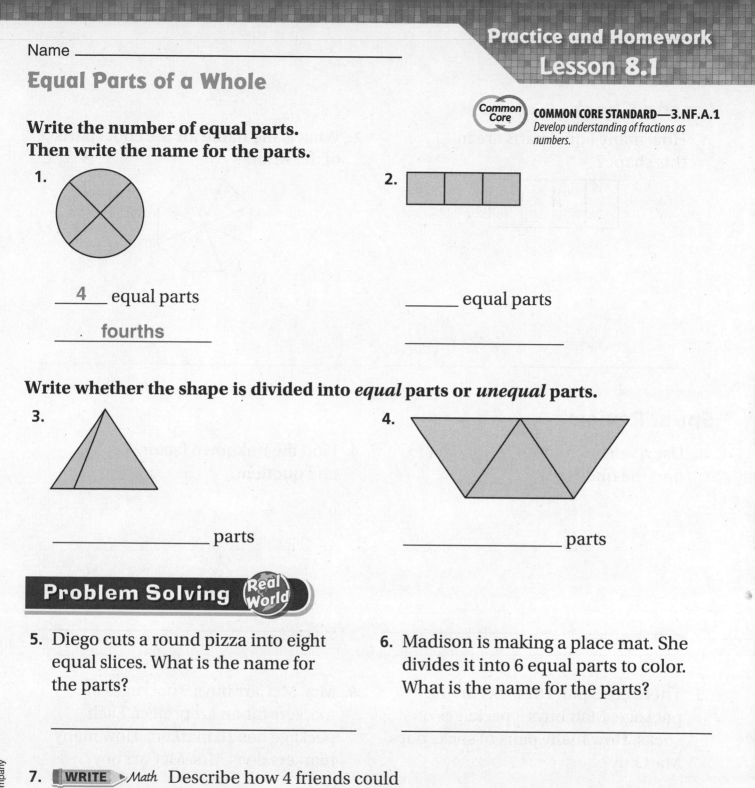

1.

___4___ equal parts

___fourths___

2.

_____ equal parts

_____

Write whether the shape is divided into *equal* parts or *unequal* parts.

3.

_____ parts

4.

_____ parts

Problem Solving · Real World

5. Diego cuts a round pizza into eight equal slices. What is the name for the parts?

_____

6. Madison is making a place mat. She divides it into 6 equal parts to color. What is the name for the parts?

_____

7. **WRITE** ▸ *Math* Describe how 4 friends could share a sandwich equally.

_____

_____

1. How many equal parts are in this shape?

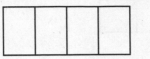

2. What is the name for the equal parts of the whole?

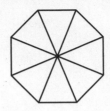

_____

_____

Spiral Review (3.OA.A.3, 3.OA.C.7)

3. Use a related multiplication fact to find the quotient.

$$49 \div 7 =$$

4. Find the unknown factor and quotient.

$$9 \times \boxed{} = 45$$

$$45 \div 9 = \boxed{}$$

_____

_____

5. There are 5 pairs of socks in one package. Matt buys 3 packages of socks. How many pairs of socks does Matt buy?

6. Mrs. McCarr buys 9 packages of markers for an art project. Each package has 10 markers. How many markers does Mrs. McCarr buy?

_____

_____

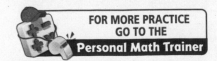

FOR MORE PRACTICE
GO TO THE
Personal Math Trainer

Name _____

Equal Shares

Essential Question Why do you need to know how to make equal shares?

Common Core · **Number and Operations— Fractions—3.NF.A.1** *Also 3.G.A.2*

MATHEMATICAL PRACTICES
MP1, MP4, MP7

Unlock the Problem Real World

Four friends share 2 small pizzas equally. What are two ways the pizza could be divided equally? How much pizza will each friend get?

🔑 **Draw to model the problem.**

Draw 2 circles to show the pizzas.

• How might the two ways be different?

_____

_____

🔑 One Way

There are _____ friends.

So, divide each pizza into 4 slices.

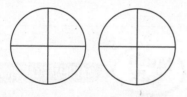

There are _____ equal parts.

Each friend can have 2 equal parts. Each friend will get 2 eighths of all the pizza.

🔑 Another Way

There are _____ friends.

So, divide all the pizza into 4 slices.

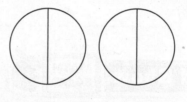

There are _____ equal parts.

Each friend can have 1 equal part. Each friend will get 1 half of a pizza.

Math Talk

MATHEMATICAL PRACTICES ❷

Use Reasoning Why does dividing the pizza into different size slices still allow the friends to have an equal share?

Try This! Four girls share 3 oranges equally. Draw a quick picture to find out how much each girl gets.

• Draw 3 circles to show the oranges.

• Draw lines to divide the circles equally.

• Shade the part 1 girl gets.

• Describe what part of an orange each girl gets.

_____

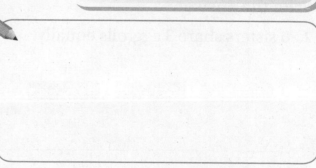

🔑 Example

Melissa and Kyle are planning to share one pan of lasagna with 6 friends. They do not agree on the way to cut the pan into equal parts. Will each friend get an equal share using Melissa's way? Using Kyle's way?

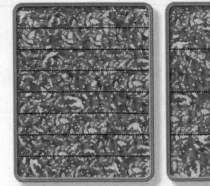

Melissa's Way   **Kyle's Way**

- Will Melissa's shares and Kyle's shares have the same shape? _____

- Will their shares using either way be the same size? _____

So, each friend will get an _____ share using either way.

- Explain why both ways let the friends have the same amount.

_____

_____

Share and Show MATH BOARD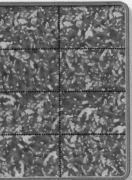

Math Talk **MATHEMATICAL PRACTICES ❻**

Explain another way the oranges could have been divided. Tell how much each friend will get.

1. Two friends share 4 oranges equally. Use the picture to find how much each friend gets.

Think: There are more oranges than friends.

_____

Draw lines to show how much each person gets.
Write the answer.

✔ 2. 8 sisters share 3 eggrolls equally.

_____

✔ 3. 6 students share 4 bagels equally.

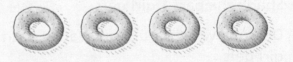

_____

Name _____

Draw lines to show how much each person gets. Write the answer.

4. 3 classmates share 2 granola bars equally.

_____

5. 4 brothers share 2 sandwiches equally.

_____

Draw to show how much each person gets. Shade the amount that one person gets. Write the answer.

6. 8 friends share 4 sheets of construction paper equally.

_____

7. **MATHEMATICAL PRACTICE ④ Model Mathematics** 4 sisters share 3 muffins equally.

_____

8. **GO DEEPER** Maria prepared 5 quesadillas. She wants to share them equally among 8 of her neighbors. How much of a quesadilla will each neighbor get?

_____

Unlock the Problem (Real World)

9. THINK SMARTER Julia holds a bread-baking class. She has 4 adults and 3 children in the class. The class will make 2 round loaves of bread. If Julia plans to give each person, including herself, an equal part of the baked breads, how much bread will each person get?

a. What do you need to find? _____

_____

b. How will you use what you know about drawing equal

shares to solve the problem? _____

_____

c. Draw a quick picture to find the share
   of bread each person will get.

d. So, each person will get

   _____ of a loaf of bread.

10. THINK SMARTER Lara and three girl friends share three sandwiches equally.

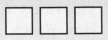

How much does each girl get? Mark all that apply.

(A) 3 fifths of a sandwich   (C) 1 whole sandwich

(B) 3 fourths of a sandwich   (D) one half and 1 fourth of a sandwich

Name _____

Equal Shares

Common Core   COMMON CORE STANDARD—3.NF.A.1
Develop understanding of fractions as numbers.

Draw lines to show how much each person gets. Write the answer.

1. 6 friends share 3 sandwiches equally.

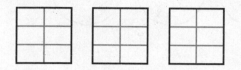

3 sixths of a sandwich _____

_____

2. 4 teammates share 5 granola bars equally.
Draw to show how much each person gets.
Shade the amount that one person gets.
Write the answer.

_____

Problem Solving • Real World

3. Three brothers share 2 sandwiches
equally. How much of a sandwich
does each brother get?

_____

_____

4. Six neighbors share 4 pies equally.
How much of a pie does each
neighbor get?

_____

_____

5. **WRITE ▸ Math** Draw a diagram to show 3 pizzas
shared equally among 6 friends.

_____

_____

Chapter 8   453

Lesson Check (3.NF.A.1)

1. Two friends share 3 fruit bars equally. How much does each friend get?

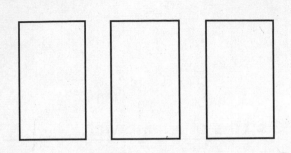

2. Four brothers share 3 pizzas equally. How much of a pizza does each brother get?

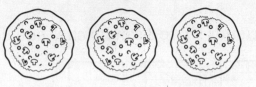

_____

_____

Spiral Review (3.OA.A.3, 3.OA.C.7, 3.NBT.A.2)

3. Find the quotient.

$$3\overline{)27}$$

4. Tyrice put 4 cookies in each of 7 bags. How many cookies in all did he put in the bags?

_____

_____

5. Ryan earned $5 per hour raking leaves. He earned $35. How many hours did he rake leaves?

6. Hannah has 229 horse stickers and 164 kitten stickers. How many more horse stickers than kitten stickers does Hannah have?

_____

_____

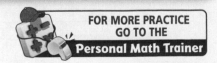

FOR MORE PRACTICE GO TO THE Personal Math Trainer

Unit Fractions of a Whole

Essential Question What do the top and bottom numbers of a fraction tell?

Common Core   **Number and Operations— Fractions—3.NF.A.1** *Also 3.G.A.2*
MATHEMATICAL PRACTICES
MP2, MP4, MP7

A **fraction** is a number that names part of a whole or part of a group.

In a fraction, the top number tells how many equal parts are being counted. $\longrightarrow$ $\dfrac{1}{6}$

The bottom number tells how many equal parts are in the whole or in the group. $\longrightarrow$

A **unit fraction** names 1 equal part of a whole. It has 1 as its top number. $\frac{1}{6}$ is a unit fraction.

Unlock the Problem   Real World

Luke's family picked strawberries. They put the washed strawberries in one part of a fruit platter. The platter had 6 equal parts. What fraction of the fruit platter had strawberries?

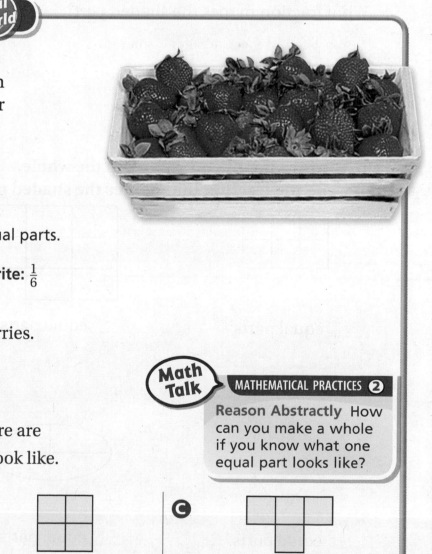

Find part of a whole.

Shade 1 of the 6 equal parts.

Read: one sixth   **Write:** $\frac{1}{6}$

So, _____ of the platter had strawberries.

Use a fraction to find a whole.

This shape ☐ is $\frac{1}{4}$ of the whole. Here are examples of what the whole could look like.

Math Talk   MATHEMATICAL PRACTICES ❷

Reason Abstractly How can you make a whole if you know what one equal part looks like?

Ⓐ   Ⓑ   Ⓒ

Try This! Look again at the examples at the bottom of page 455.
Draw two other pictures of how the whole might look.

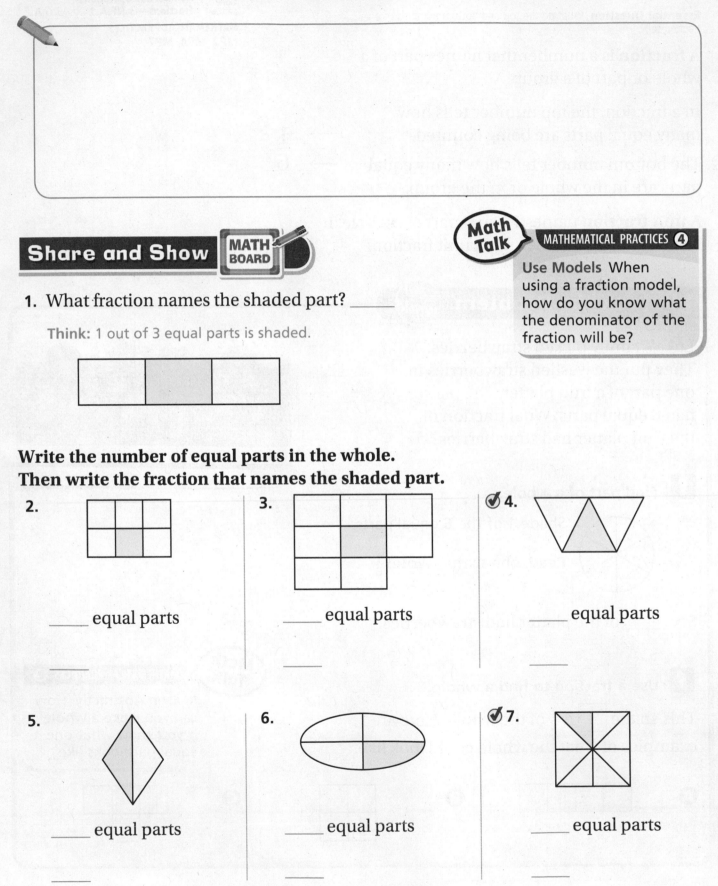

Share and Show MATH BOARD

Math Talk **MATHEMATICAL PRACTICES** ④

Use Models When using a fraction model, how do you know what the denominator of the fraction will be?

1. What fraction names the shaded part? _____

   Think: 1 out of 3 equal parts is shaded.

**Write the number of equal parts in the whole.
Then write the fraction that names the shaded part.**

2. _____ equal parts

   _____

3. _____ equal parts

   _____

✓ 4. _____ equal parts

   _____

5. _____ equal parts

   _____

6. _____ equal parts

   _____

✓ 7. _____ equal parts

   _____

© Houghton Mifflin Harcourt Publishing Company

On Your Own

Write the number of equal parts in the whole.
Then write the fraction that names the shaded part.

8.

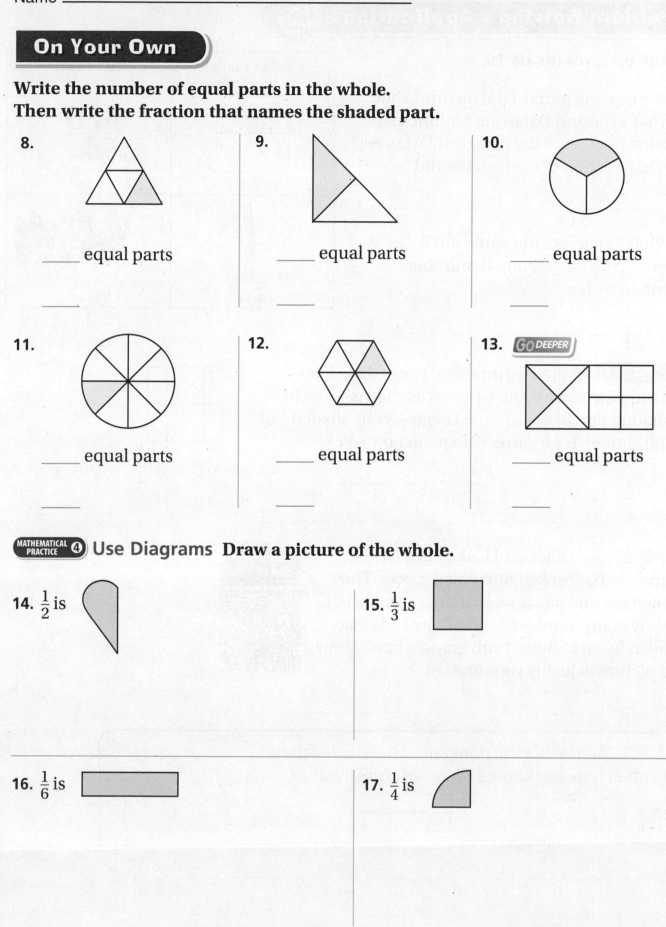

_____ equal parts

_____

9.

_____ equal parts

_____

10.

_____ equal parts

_____

11.

_____ equal parts

_____

12.

_____ equal parts

_____

13. GO DEEPER

_____ equal parts

_____

MATHEMATICAL PRACTICE ④ Use Diagrams **Draw a picture of the whole.**

14. $\frac{1}{2}$ is

15. $\frac{1}{3}$ is

16. $\frac{1}{6}$ is

17. $\frac{1}{4}$ is

Problem Solving • Applications

Use the pictures for 18–19.

Kylie's Lunch	Dylan's Lunch
sandwich	pizza
apple	fruit bar

18. The missing parts of the pictures show what Kylie and Dylan ate for lunch. What fraction of the pizza did Dylan eat? What fraction of the fruit bar did he eat?

_____

19. What fraction of the apple did Kylie eat? Write the fraction in numbers and in words.

____   _____

20. **MATHEMATICAL PRACTICE 3 Make Arguments** Diego drew lines to divide the square into 6 pieces as shown. Then he shaded part of the square. Diego says he shaded $\frac{1}{6}$ of the square. Is he correct? Explain how you know.

_____

_____

21. **THINK SMARTER** Riley and Chad each have a granola bar broken into equal pieces. They each eat one piece, or $\frac{1}{4}$, of their granola bar. How many more pieces do Riley and Chad need to eat to finish both granola bars? Draw a picture to justify your answer.

_____

22. **THINK SMARTER** What fraction names the shaded part? Explain how you know how to write the fraction.

_____

Name _____

Unit Fractions of a Whole

Common Core   **COMMON CORE STANDARD—3.NF.A.1**
Develop understanding of fractions as numbers.

**Write the number of equal parts in the whole.
Then write the fraction that names the shaded part.**

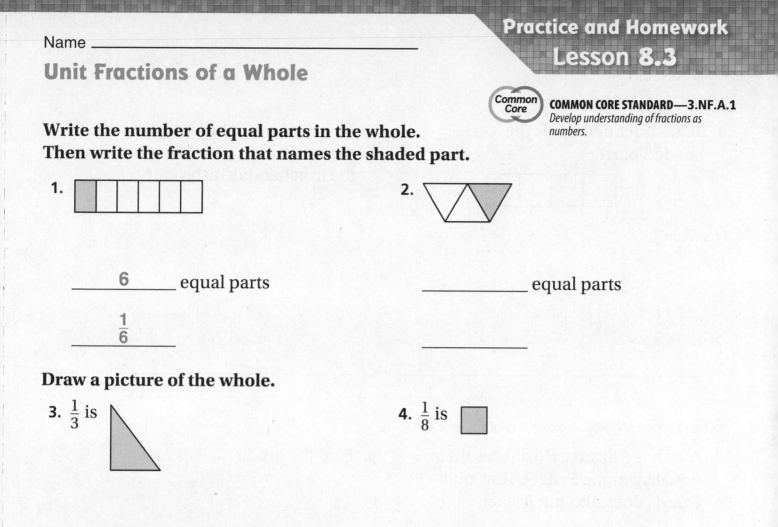

1.

_____6_____ equal parts

____1/6____

2.

_____ equal parts

_____

Draw a picture of the whole.

3. $\frac{1}{3}$ is

4. $\frac{1}{8}$ is

Problem Solving · Real World

5. Tyler made a pan of cornbread. He cut it into 8 equal pieces and ate 1 piece. What fraction of the cornbread did Tyler eat?

_____

6. Anna cut an apple into 4 equal pieces. She gave 1 piece to her sister. What fraction of the apple did Anna give to her sister?

_____

7. **WRITE** ▸*Math*   Draw a picture to show what 1 out of 3 equal parts looks like. Then write the fraction.

_____

_____

Lesson Check (3.NF.A.1)

1. What fraction names the shaded part?

   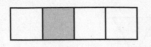

   _____

2. Tasha cut a fruit bar into 3 equal parts. She ate 1 part. What fraction of the fruit bar did Tasha eat?

   _____

Spiral Review (3.OA.A.3, 3.OA.B.5, 3.MD.B.3)

3. Alex has 5 lizards. He divides them equally among 5 cages. How many lizards does Alex put in each cage?

   _____

4. Find the product.

   $$8 \times 1 = \boxed{}$$

   _____

5. Leo bought 6 chew toys for his new puppy. Each chew toy cost $4. How much did Leo spend for the chew toys?

   _____

6. Lilly is making a picture graph. Each picture of a star is equal to two books she has read. The row for the month of December has 3 stars. How many books did Lilly read during the month of December?

   _____

© Houghton Mifflin Harcourt Publishing Company

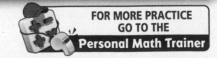

FOR MORE PRACTICE
GO TO THE
Personal Math Trainer

Name _____

Fractions of a Whole

Essential Question How does a fraction name part of a whole?

Common Core **Number and Operations—Fractions—3.NF.A.1** *Also 3.G.A.2*
MATHEMATICAL PRACTICES
MP2, MP4, MP7

🔑 Unlock the Problem Real World

The first pizzeria in America opened in New York in 1905. The pizza recipe came from Italy. Look at Italy's flag. What fraction of the flag is not red?

 Name equal parts of a whole.

A fraction can name more than 1 equal part of a whole.

The flag is divided into 3 equal parts, and 2 parts are not red.

2 parts not red → $\frac{2}{3}$ ← numerator
3 equal parts in all   ← denominator

Read: two thirds or two parts out of three equal parts

Write: $\frac{2}{3}$

So, _____ of the flag is not red.

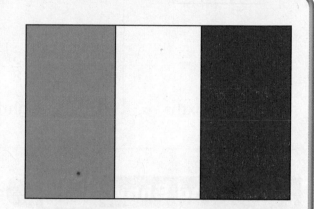
▲ Italy's flag has three equal parts.

Math Idea
When all the parts are shaded, one whole shape is equal to all of its parts. It represents the whole number 1.
$\frac{3}{3} = 1$

The **numerator** tells how many parts are being counted.

The **denominator** tells how many equal parts are in the whole or in the group.

You can count equal parts, such as sixths, to make a whole.

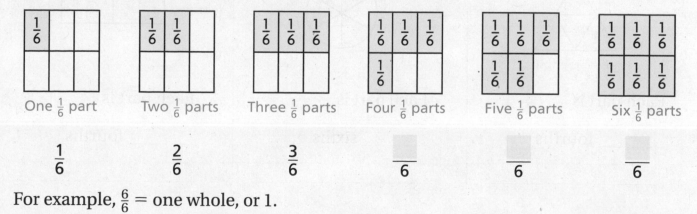

For example, $\frac{6}{6}$ = one whole, or 1.

Try This! Write the missing word or number to name the shaded part.

A

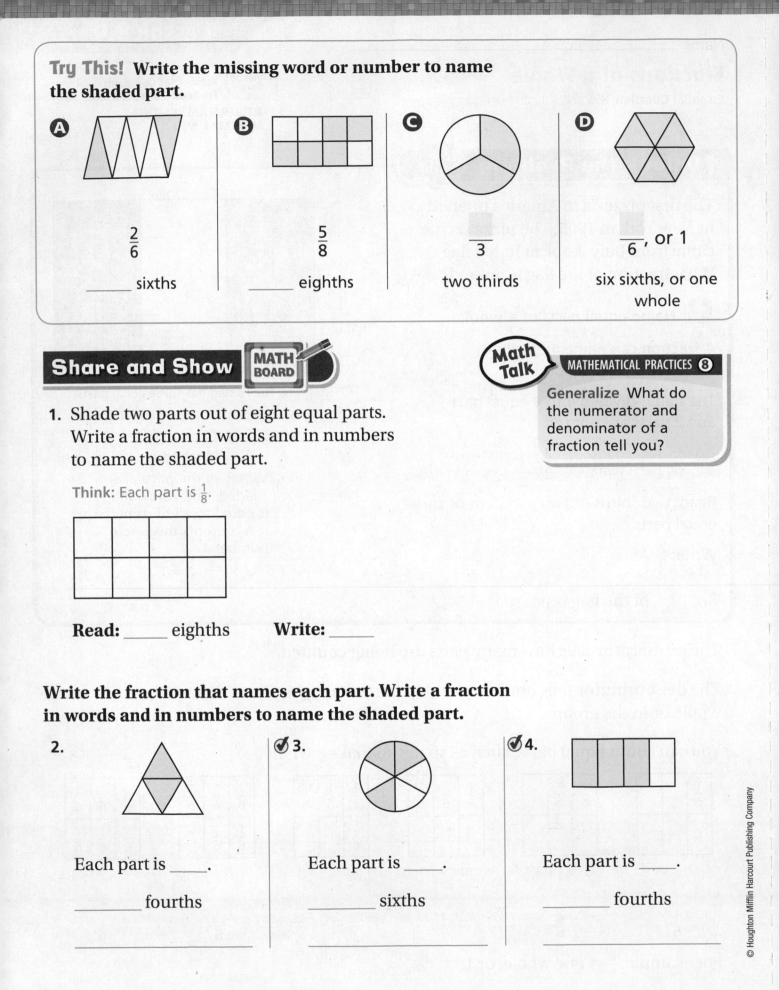

$\dfrac{2}{6}$

_____ sixths

B

$\dfrac{5}{8}$

_____ eighths

C

$\dfrac{}{3}$

two thirds

D

$\dfrac{}{6}$, or 1

six sixths, or one whole

Share and Show MATH BOARD

Math Talk MATHEMATICAL PRACTICES ⑧

Generalize What do the numerator and denominator of a fraction tell you?

1. Shade two parts out of eight equal parts. Write a fraction in words and in numbers to name the shaded part.

   Think: Each part is $\dfrac{1}{8}$.

   Read: _____ eighths   **Write:** _____

Write the fraction that names each part. Write a fraction in words and in numbers to name the shaded part.

2.

Each part is _____.

_____ fourths

_____

☑ 3.

Each part is _____.

_____ sixths

_____

☑ 4.

Each part is _____.

_____ fourths

_____

462

© Houghton Mifflin Harcourt Publishing Company

On Your Own

Write the fraction that names each part. Write a fraction in words and in numbers to name the shaded part.

5.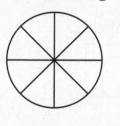

Each part is ____.

_____ eighths

_____

6.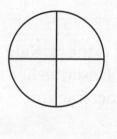

Each part is ____.

_____ thirds

_____

7.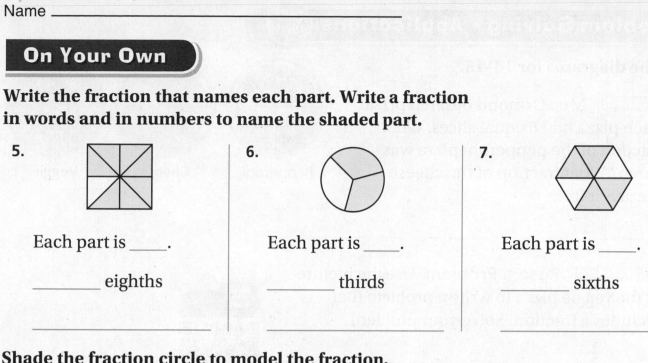

Each part is ____.

_____ sixths

_____

Shade the fraction circle to model the fraction. Then write the fraction in numbers.

8. six out of eight

_____

9. three fourths

_____

10. three out of three

_____

11. A flag is divided into four equal sections. One section is white. What fraction of the flag is not white?

_____

12. A garden has six sections. Two sections are planted with tomatoes. Which fraction represents the part of the garden without tomatoes?

_____

13. Jane is making a memory quilt from some of her old favorite clothes that are too small. She will use T-shirts for the shaded squares in the pattern. What names the part of the quilt that will be made of T-shirts?

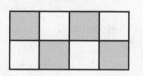

_____

Problem Solving • Applications

Use the diagrams for 14–15.

Pepperoni   Cheese   Veggie

14. GoDEEPER Mrs. Ormond ordered pizza. Each pizza had 8 equal slices. What fraction of the pepperoni pizza was eaten? What fraction of the cheese pizza is left?

_____

15. THINKSMARTER **Pose a Problem** Use the picture of the veggie pizza to write a problem that includes a fraction. Solve your problem.

_____

_____

_____

16. MATHEMATICAL PRACTICE ③ **Verify the Reasoning of Others** Kate says that $\frac{2}{4}$ of the rectangle is shaded. Describe her error. Use the model to write the correct fraction for the shaded part.

_____

_____

_____

17. THINKSMARTER Select a numerator and a denominator for the fraction that names the shaded part of the shape.

Numerator	Denominator
○ 2	○ 3
○ 3	○ 5
○ 5	○ 6
○ 6	○ 8

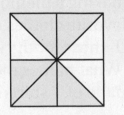

464

Name _____

Fractions of a Whole

Write the fraction that names each part. Write a fraction in words and in numbers to name the shaded part.

Common Core

COMMON CORE STANDARD—3.NF.A.1
Develop understanding of fractions as numbers.

1.

Each part is ____$\frac{1}{6}$____.

___three___ sixths

___$\frac{3}{6}$___

2.

Each part is _____.

_____ eighths

_____

Shade the fraction circle to model the fraction. Then write the fraction in numbers.

3. four out of six

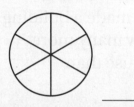

_____

4. eight out of eight

_____

Problem Solving · Real World

5. Emma makes a poster for the school's spring concert. She divides the poster into 8 equal parts. She uses two of the parts for the title. What fraction of the poster does Emma use for the title?

_____

6. Lucas makes a flag. It has 6 equal parts. Five of the parts are red. What fraction of the flag is red?

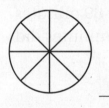

_____

7. **WRITE** ▸*Math* Draw a rectangle and divide it into 4 equal parts. Shade 3 parts. Then write the fraction that names the shaded part.

_____

_____

Lesson Check (3.NF.A.1)

1. What fraction names the shaded part?

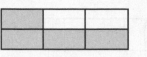

2. What fraction names the shaded part?

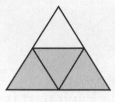

Spiral Review (3.OA.C.7, 3.NBT.A.2, 3.MD.B.3)

3. Sarah biked for 115 minutes last week. Jennie biked for 89 minutes last week. How many minutes did the girls bike?

4. Harrison made a building using 124 blocks. Greyson made a building using 78 blocks. How many more blocks did Harrison use than Greyson did?

5. Von buys a bag of 24 dog treats. He gives his puppy 3 treats a day. How many days will the bag of dog treats last?

6. How many students chose swimming?

Favorite Activity	
Skating	☺ ☺
Swimming	☺ ☺ ☺ ☺ ☺
Biking	☺ ☺ ☺ ☺
Key: Each ☺ = 5 votes.	

466

FOR MORE PRACTICE GO TO THE
Personal Math Trainer

Fractions on a Number Line

Essential Question How can you represent and locate fractions on a number line?

Common Core

**Number and Operations—
Fractions—3.NF.A.2a, 3.NF.A.2b**
Also 3.NF.A.2

MATHEMATICAL PRACTICES
MP1, MP4, MP7

🔑 Unlock the Problem Real World

Billy's family is traveling from his house to his grandma's house. They stop at gas stations when they are $\frac{1}{4}$ and $\frac{3}{4}$ of the way there. How can you represent those distances on a number line?

You can use a number line to show fractions. The length from one whole number to the next whole number represents one whole. The line can be divided into any number of equal parts, or lengths.

> **Math Idea**
>
> A point on a number line shows the endpoint of a length, or distance, from zero. A number or fraction can name the distance.

🔓 Activity Locate fractions on a number line.

Materials ■ fraction strips

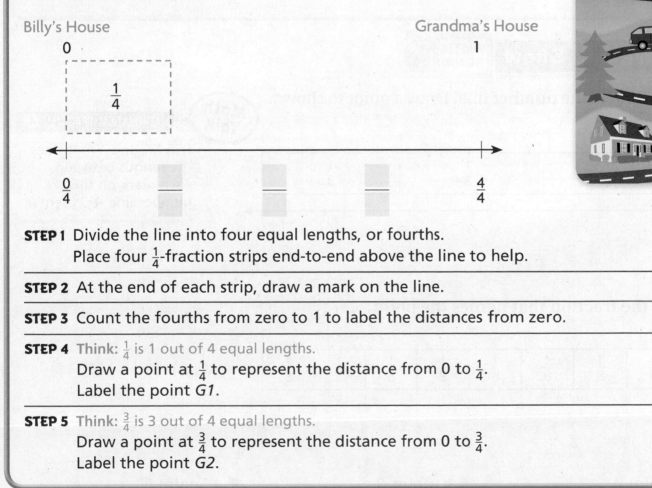

Billy's House

0

$\frac{1}{4}$

$\frac{0}{4}$ $\frac{4}{4}$

Grandma's House

1

STEP 1 Divide the line into four equal lengths, or fourths.
Place four $\frac{1}{4}$-fraction strips end-to-end above the line to help.

STEP 2 At the end of each strip, draw a mark on the line.

STEP 3 Count the fourths from zero to 1 to label the distances from zero.

STEP 4 Think: $\frac{1}{4}$ is 1 out of 4 equal lengths.
Draw a point at $\frac{1}{4}$ to represent the distance from 0 to $\frac{1}{4}$.
Label the point *G1*.

STEP 5 Think: $\frac{3}{4}$ is 3 out of 4 equal lengths.
Draw a point at $\frac{3}{4}$ to represent the distance from 0 to $\frac{3}{4}$.
Label the point *G2*.

❶ Example Complete the number line to name the point.

Materials ■ color pencils

Write the fraction that names the point on the number line.

Think: This number line is divided into six equal lengths, or sixths.

The length of one equal part is _____.

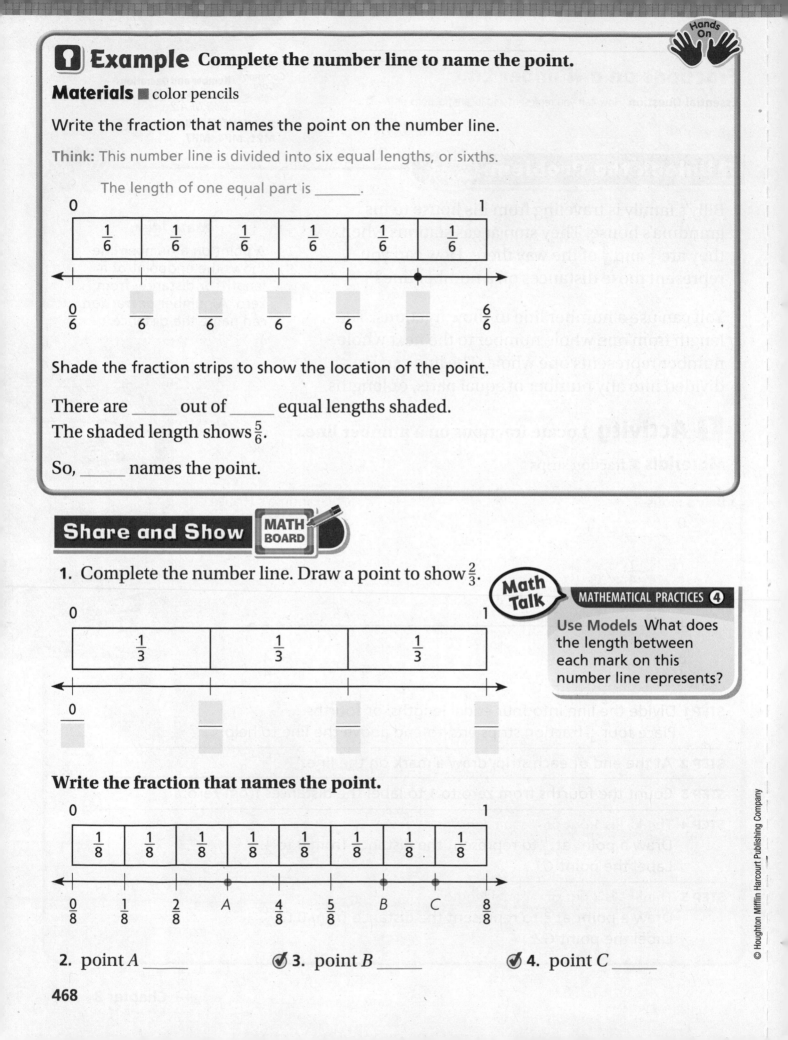

0   1

| $\frac{1}{6}$ | $\frac{1}{6}$ | $\frac{1}{6}$ | $\frac{1}{6}$ | $\frac{1}{6}$ | $\frac{1}{6}$ |

$\frac{0}{6}$ $\frac{}{6}$ $\frac{}{6}$ $\frac{}{6}$ $\frac{}{6}$ $\frac{6}{6}$

Shade the fraction strips to show the location of the point.

There are _____ out of _____ equal lengths shaded.
The shaded length shows $\frac{5}{6}$.

So, _____ names the point.

Share and Show MATH BOARD

1. Complete the number line. Draw a point to show $\frac{2}{3}$.

Math Talk

MATHEMATICAL PRACTICES ❹

Use Models What does the length between each mark on this number line represents?

0   1

| $\frac{1}{3}$ | $\frac{1}{3}$ | $\frac{1}{3}$ |

$\frac{0}{}$

Write the fraction that names the point.

0   1

| $\frac{1}{8}$ | $\frac{1}{8}$ | $\frac{1}{8}$ | $\frac{1}{8}$ | $\frac{1}{8}$ | $\frac{1}{8}$ | $\frac{1}{8}$ | $\frac{1}{8}$ |

$\frac{0}{8}$ $\frac{1}{8}$ $\frac{2}{8}$ A $\frac{4}{8}$ $\frac{5}{8}$ B   C $\frac{8}{8}$

2. point A _____   ✓ **3.** point B _____   ✓ **4.** point C _____

On Your Own

Use fraction strips to help you complete the number line. Then locate and draw a point for the fraction.

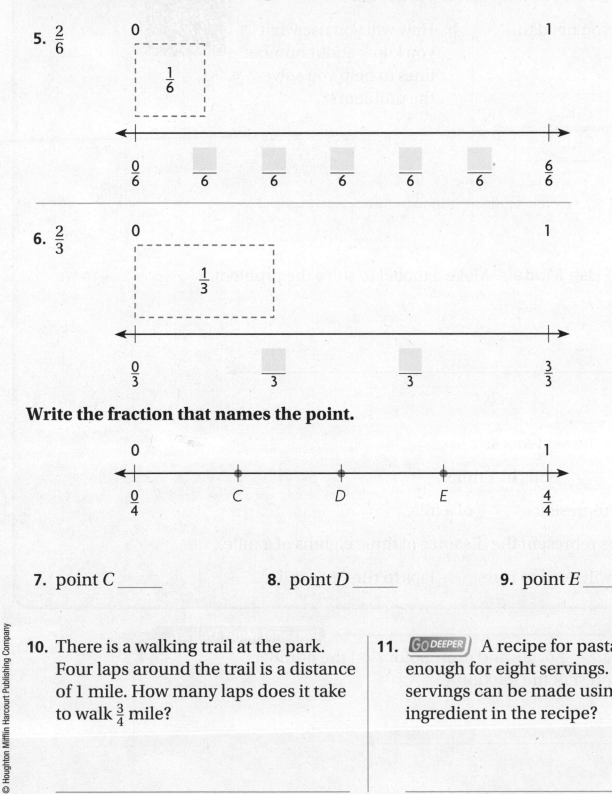

5. $\frac{2}{6}$

0   1

$\frac{1}{6}$

$\frac{0}{6}$ $\frac{}{6}$ $\frac{}{6}$ $\frac{}{6}$ $\frac{}{6}$ $\frac{}{6}$ $\frac{6}{6}$

6. $\frac{2}{3}$

0   1

$\frac{1}{3}$

$\frac{0}{3}$ $\frac{}{3}$ $\frac{}{3}$ $\frac{3}{3}$

Write the fraction that names the point.

0   1

$\frac{0}{4}$ C   D   E $\frac{4}{4}$

7. point C _____

8. point D _____

9. point E _____

10. There is a walking trail at the park. Four laps around the trail is a distance of 1 mile. How many laps does it take to walk $\frac{3}{4}$ mile?

11. **GO DEEPER** A recipe for pasta makes enough for eight servings. How many servings can be made using $\frac{4}{8}$ of each ingredient in the recipe?

🔑 Unlock the Problem (Real World)

12. **THINK SMARTER** Javia ran 8 laps around a track to run a total of 1 mile on Monday. How many laps will she need to run on Tuesday to run $\frac{3}{8}$ of a mile?

a. What do you need to find?

_____

_____

_____

b. How will you use what you know about number lines to help you solve the problem?

_____

_____

_____

_____

c. **MATHEMATICAL PRACTICE ④** **Use Models** Make a model to solve the problem.

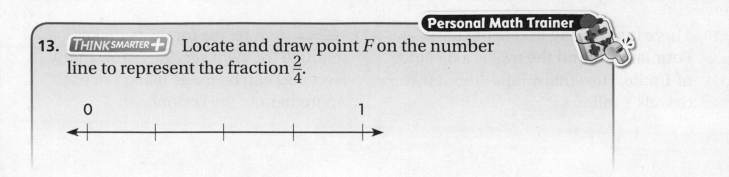

d. Complete the sentences.

There are _____ laps in 1 mile.

Each lap represents _____ of a mile.

_____ laps represent the distance of three eighths of a mile.

So, Javia will need to run _____ laps to run $\frac{3}{8}$ of a mile.

Personal Math Trainer

13. **THINK SMARTER +** Locate and draw point *F* on the number line to represent the fraction $\frac{2}{4}$.

Fractions on a Number Line

Common Core

COMMON CORE STANDARDS—3.NF.A.2a, 3.NF.A.2b *Develop understanding of fractions as numbers.*

Use fraction strips to help you complete the number line.
Then locate and draw a point for the fraction.

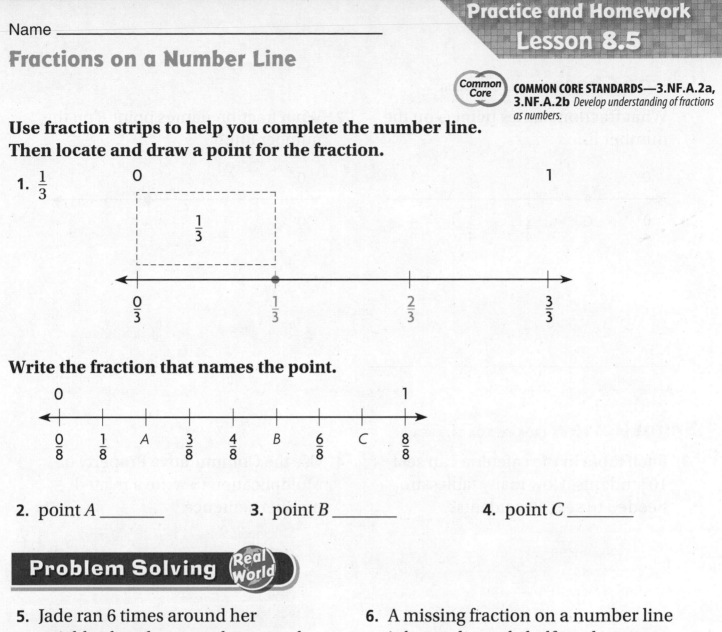

1. $\frac{1}{3}$

Write the fraction that names the point.

2. point A _____   3. point B _____   4. point C _____

5. Jade ran 6 times around her neighborhood to complete a total of 1 mile. How many times will she need to run to complete $\frac{5}{6}$ of a mile?

_____

6. A missing fraction on a number line is located exactly halfway between $\frac{3}{6}$ and $\frac{5}{6}$. What is the missing fraction?

_____

7. **WRITE** ▸*Math* Explain how showing fractions with models and a number line are alike and different.

_____

_____

Lesson Check (3.NF.A.2a, 3.NF.A.2b)

1. What fraction names point *G* on the number line?

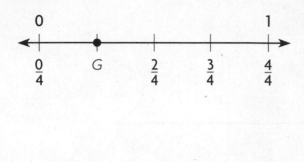

2. What fraction names point *R* on the number line?

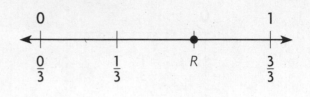

_____

_____

Spiral Review (3.OA.B.5, 3.OA.C.7, 3.NF.A.1)

3. Each table in the cafeteria can seat 10 students. How many tables are needed to seat 40 students?

4. Use the Commutative Property of Multiplication to write a related number sentence.

$$4 \times 9 = 36$$

_____

_____

5. Pedro shaded part of a circle. What fraction names the shaded part?

6. Find the quotient.

$$8 \div 1 = \boxed{}$$

_____

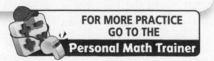

FOR MORE PRACTICE
GO TO THE
Personal Math Trainer

✔ Mid-Chapter Checkpoint

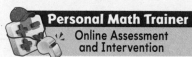

Personal Math Trainer
Online Assessment
and Intervention

Vocabulary

Choose the best term from the box to complete the sentence.

Vocabulary
denominator
fraction
numerator

1. A _____ is a number that names part of a whole or part of a group. (p. 455)

2. The _____ tells how many equal parts are in the whole or in the group. (p. 461)

Concepts and Skills

Write the number of equal parts. Then write the name for the parts. (3.NF.A.1)

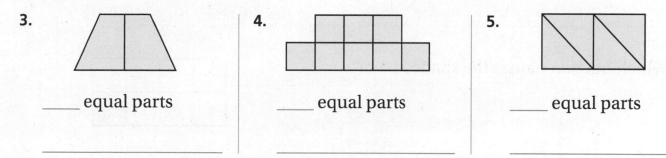

3. ____ equal parts

____

4. ____ equal parts

____

5. ____ equal parts

____

Write the number of equal parts in the whole. Then write the fraction that names the shaded part. (3.NF.A.1)

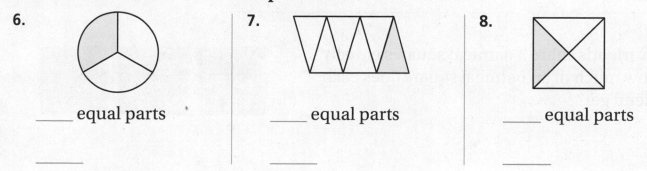

6. ____ equal parts

____

7. ____ equal parts

____

8. ____ equal parts

____

Write the fraction that names the point. (3.NF.A.2a, 3.NF.A.2b)

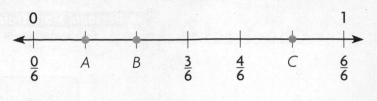

9. point A _____   10. point B _____   11. point C _____

12. **GO DEEPER** Jessica ordered a pizza. What fraction of the pizza has mushrooms? What fraction of the pizza does not have mushrooms? (3.NF.A.1)

_____

13. Which fraction names the shaded part?

   (3.NF.A.1)

14. Six friends share 3 oatmeal squares equally. How much of an oatmeal square does each friend get? (3.NF.A.1)

_____

474

Name _____

Relate Fractions and Whole Numbers

Essential Question When might you use a fraction greater than 1 or a whole number?

Common Core **Number and Operations—Fractions—3.NF.A.3c** *Also 3.NF.A.2, 3.NF.A.2b, 3.G.A.2*
MATHEMATICAL PRACTICES
MP1, MP4, MP6, MP7

🔑 Unlock the Problem · Real World

Steve ran 1 mile and Jenna ran $\frac{4}{4}$ of a mile. Did Steve and Jenna run the same distance?

🔑 **Locate 1 and $\frac{4}{4}$ on a number line.**

- Shade 4 lengths of $\frac{1}{4}$ and label the number line.

- Draw a point at 1 and $\frac{4}{4}$.

> **Math Idea**
> If two numbers are located at the same point on a number line, then they are equal and represent the same distance.

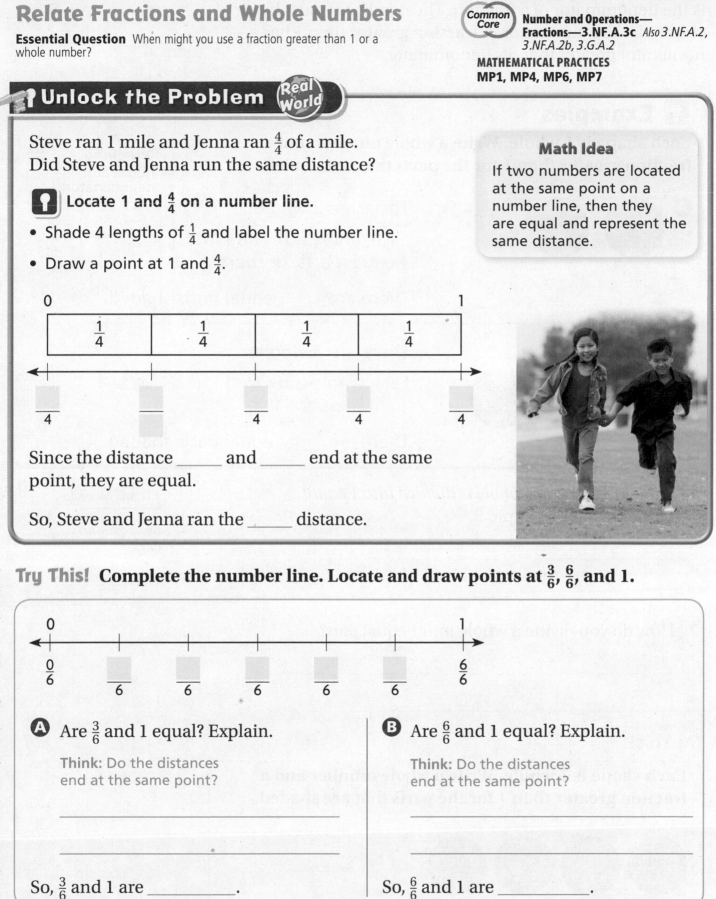

Since the distance _____ and _____ end at the same point, they are equal.

So, Steve and Jenna ran the _____ distance.

Try This! Complete the number line. Locate and draw points at $\frac{3}{6}$, $\frac{6}{6}$, and 1.

A Are $\frac{3}{6}$ and 1 equal? Explain.

Think: Do the distances end at the same point?

_____

_____

So, $\frac{3}{6}$ and 1 are _____.

B Are $\frac{6}{6}$ and 1 equal? Explain.

Think: Do the distances end at the same point?

_____

_____

So, $\frac{6}{6}$ and 1 are _____.

CONNECT The number of equal parts the whole is divided into is the denominator of a fraction. The number of parts being counted is the numerator. A **fraction greater than 1** has a numerator greater than its denominator.

🔒 Examples

Each shape is 1 whole. Write a whole number and a fraction greater than 1 for the parts that are shaded.

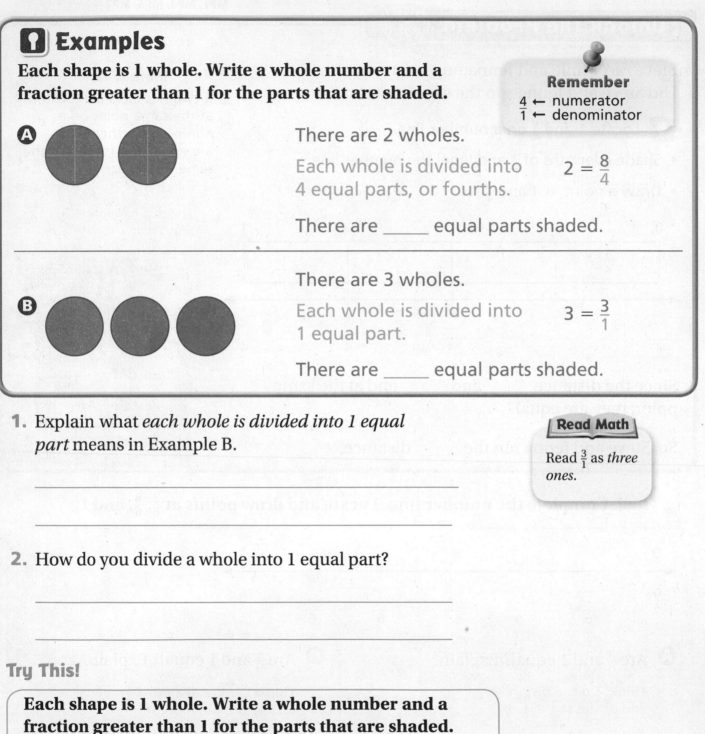

A

There are 2 wholes.

Each whole is divided into 4 equal parts, or fourths. $2 = \frac{8}{4}$

There are _____ equal parts shaded.

B

There are 3 wholes.

Each whole is divided into 1 equal part. $3 = \frac{3}{1}$

There are _____ equal parts shaded.

1. Explain what *each whole is divided into 1 equal part* means in Example B.

_____

_____

Read Math

Read $\frac{3}{1}$ as *three ones.*

2. How do you divide a whole into 1 equal part?

_____

_____

Try This!

Each shape is 1 whole. Write a whole number and a fraction greater than 1 for the parts that are shaded.

$$\square = \frac{\square}{\square}$$

Share and Show MATH BOARD

1. Each shape is 1 whole. Write a whole number and a fraction greater than 1 for the parts that are shaded.

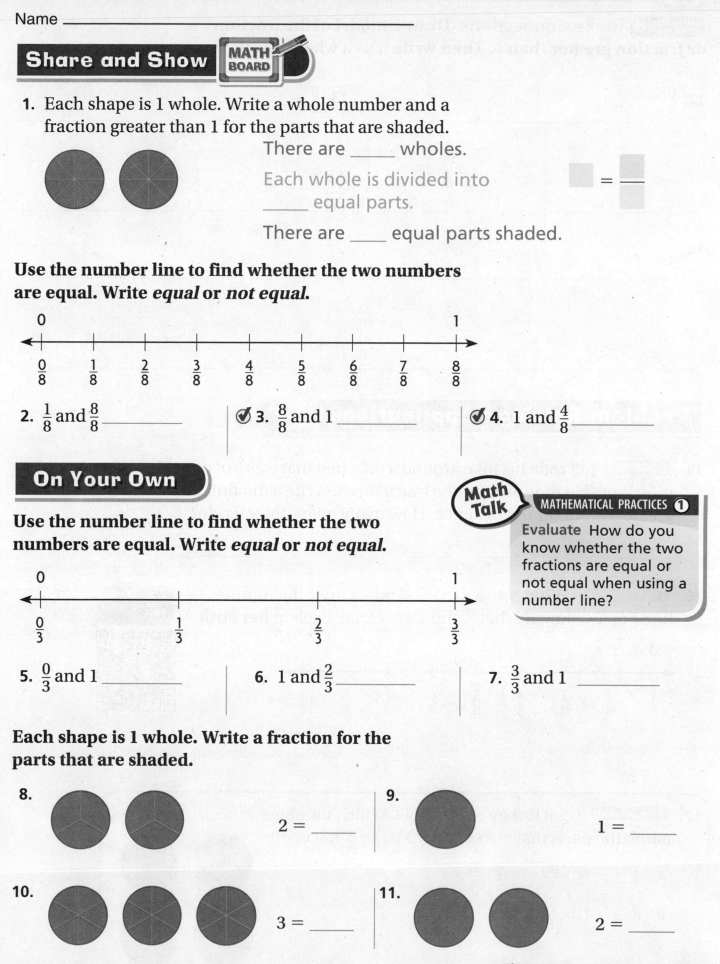

There are _____ wholes.

Each whole is divided into _____ equal parts.

There are _____ equal parts shaded.

□ = —

Use the number line to find whether the two numbers are equal. Write _equal_ or _not equal_.

0   1
⊢———⊢———⊢———⊢———⊢———⊢———⊢———⊢———⊢
$\frac{0}{8}$ $\frac{1}{8}$ $\frac{2}{8}$ $\frac{3}{8}$ $\frac{4}{8}$ $\frac{5}{8}$ $\frac{6}{8}$ $\frac{7}{8}$ $\frac{8}{8}$

2. $\frac{1}{8}$ and $\frac{8}{8}$ _____

✓ 3. $\frac{8}{8}$ and 1 _____

✓ 4. 1 and $\frac{4}{8}$ _____

On Your Own

Use the number line to find whether the two numbers are equal. Write _equal_ or _not equal_.

Math Talk MATHEMATICAL PRACTICES ①

Evaluate How do you know whether the two fractions are equal or not equal when using a number line?

0   1
⊢—————————⊢—————————⊢—————————⊢
$\frac{0}{3}$ $\frac{1}{3}$ $\frac{2}{3}$ $\frac{3}{3}$

5. $\frac{0}{3}$ and 1 _____

6. 1 and $\frac{2}{3}$ _____

7. $\frac{3}{3}$ and 1 _____

Each shape is 1 whole. Write a fraction for the parts that are shaded.

8. $2 =$ _____

9. $1 =$ _____

10. $3 =$ _____

11. $2 =$ _____

MATHEMATICAL PRACTICE ⑥ Make Connections Draw a model of the fraction or fraction greater than 1. Then write it as a whole number.

12. $\frac{8}{4}$ = _____

13. $\frac{6}{6}$ = _____

14. $\frac{5}{1}$ = _____

Problem Solving · Applications Real World

15. **GO DEEPER** Jeff rode his bike around a bike trail that was $\frac{1}{3}$ of a mile long. He rode around the trail 9 times. Write a fraction greater than 1 for the distance. How many miles did Jeff ride?

_____

16. **THINK SMARTER** **What's the Error?** Andrea drew the number line below. She said that $\frac{9}{8}$ and 1 are equal. Explain her error.

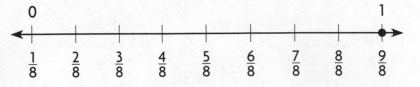

_____

17. **THINK SMARTER** Each shape is 1 whole. Which numbers name the parts that are shaded? Mark all that apply.

(A) 4   (C) $\frac{26}{6}$ (E) $\frac{6}{4}$

(B) 6   (D) $\frac{24}{6}$

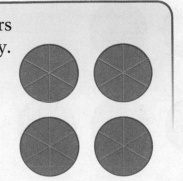

Relate Fractions and Whole Numbers

Use the number line to find whether the two numbers are equal. Write *equal* or *not equal*.

Common Core

COMMON CORE STANDARD—3.NF.A.3c
Develop an understanding of fractions as numbers.

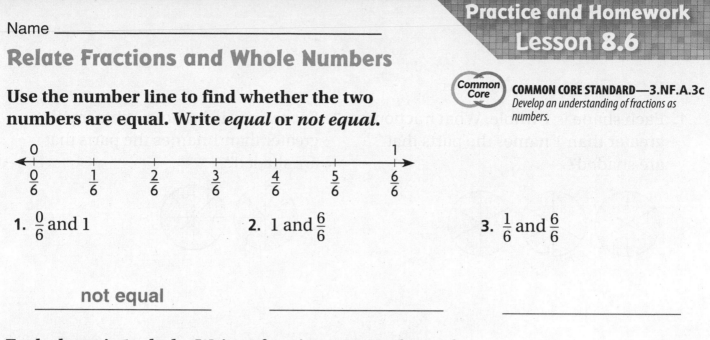

1. $\frac{0}{6}$ and 1

2. 1 and $\frac{6}{6}$

3. $\frac{1}{6}$ and $\frac{6}{6}$

_____ not equal _____

_____

_____

Each shape is 1 whole. Write a fraction greater than 1 for the parts that are shaded.

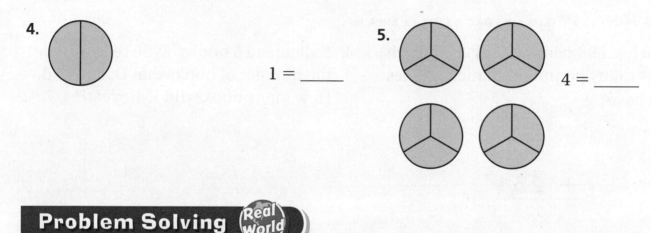

4.

5.

$1 =$ _____

$4 =$ _____

Problem Solving · Real World

6. Rachel jogged along a trail that was $\frac{1}{4}$ of a mile long. She jogged along the trail 8 times. How many miles did Rachel jog?

7. Jon ran around a track that was $\frac{1}{8}$ of a mile long. He ran around the track 24 times. How many miles did Jon run?

_____

_____

8. **WRITE** ▸*Math* Write a problem that uses a fraction greater than 1.

_____

_____

Lesson Check (3.NF.A.3c)

1. Each shape is 1 whole. What fraction greater than 1 names the parts that are shaded?

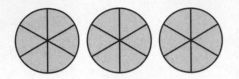

_____

2. Each shape is 1 whole. What fraction greater than 1 names the parts that are shaded?

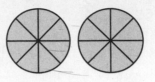

_____

Spiral Review (3.OA.A.3, 3.OA.C.7, 3.NBT.A.2, 3.NF.A.1)

3. Tara has 598 pennies and 231 nickels. How many pennies and nickels does she have?

$$598 + 231$$

_____

4. Dylan read 6 books. Kylie read double the number of books that Dylan read. How many books did Kylie read?

_____

5. Alyssa divides a granola bar into halves. How many equal parts are there?

_____

6. There are 4 students in each small reading group. If there are 24 students in all, how many reading groups are there?

_____

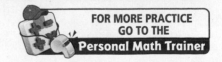

FOR MORE PRACTICE
GO TO THE
Personal Math Trainer

Name _____

Fractions of a Group

Essential Question How can a fraction name part of a group?

Common Core **Number and Operations—Fractions—3.NF.A.1**

MATHEMATICAL PRACTICES
MP1, MP4, MP5

Unlock the Problem Real World

Jake and Emma each have a collection of marbles.
What fraction of each collection is blue?

🔑 You can use a fraction to name part of a group.

Jake's Marbles	Emma's Marbles
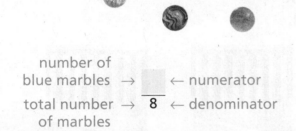	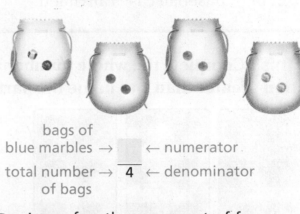

number of blue marbles → ☐ ← numerator
total number of marbles → 8 ← denominator

bags of blue marbles → ☐ ← numerator
total number of bags → 4 ← denominator

Read: three eighths, or three out of eight

Write: $\frac{3}{8}$

Read: one fourth, or one out of four

Write: $\frac{1}{4}$

So, _____ of Jake's marbles are blue.

So, _____ of Emma's marbles are blue.

Try This! **Name part of a group.**

Draw 2 red counters and 6 yellow counters.

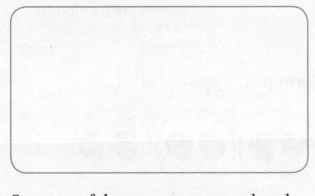

Write the fraction of counters that are red.

☐ ← number of red counters
—
☐ ← total number of counters

Write the fraction of counters that are not red.

☐ ← number of yellow counters
—
☐ ← total number of counters

So, _____ of the counters are red and _____ are not red.

Fractions Greater Than 1

Sometimes a fraction can name more than a whole group.

Daniel collects baseballs. He has collected 8 so far. He puts them in cases that hold 4 baseballs each. What part of the baseball cases has Daniel filled?

Think: 1 case = 1

Daniel has two full cases of 4 baseballs each.

So, 2, or $\frac{8}{4}$, baseball cases are filled.

Try This! Complete the whole number and the fraction greater than 1 to name the part filled.

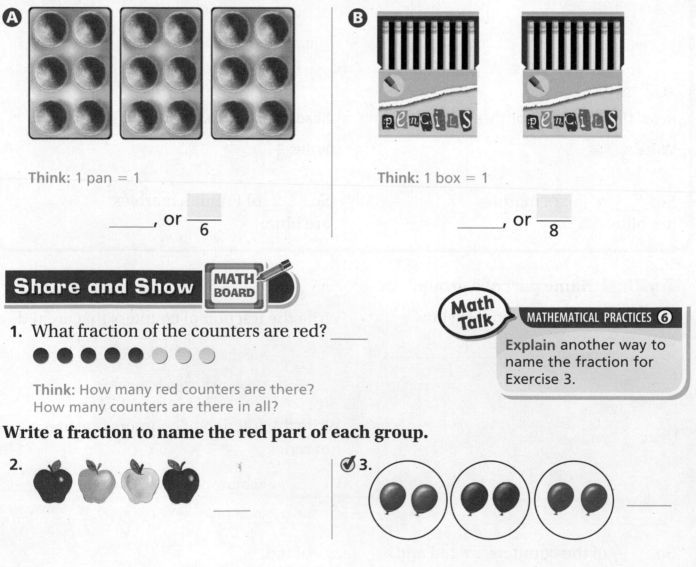

A

Think: 1 pan = 1

_____ , or $\dfrac{\quad}{6}$

B

Think: 1 box = 1

_____ , or $\dfrac{\quad}{8}$

Share and Show MATH BOARD

1. What fraction of the counters are red? _____

● ● ● ● ● ● ○ ○ ○

Think: How many red counters are there? How many counters are there in all?

Write a fraction to name the red part of each group.

2.

3.

Math Talk

MATHEMATICAL PRACTICES ⑥

Explain another way to name the fraction for Exercise 3.

Name _____

Write a whole number and a fraction greater than 1 to name the part filled.

4.

Think: 1 carton = 1

_____ _____

5.

Think: 1 container = 1

_____ _____

On Your Own

Write a fraction to name the blue part of each group.

6.

_____

7.

_____

8.

_____

9.

_____

Write a whole number and a fraction greater than 1 to name the part filled.

10.

Think: 1 container = 1

_____ _____

11. THINK SMARTER

Think: 1 carton = 1

_____ _____

Draw a quick picture on your MathBoard. Then write a fraction to name the shaded part of the group.

12. Draw 8 circles.
Shade 8 circles.

_____

13. Draw 8 triangles.
Make 4 groups.
Shade 1 group.

_____

14. Draw 4 rectangles.
Shade 2 rectangles.

_____

Problem Solving · Applications Real World

Use the graph for 15–16.

School Marble Tournament

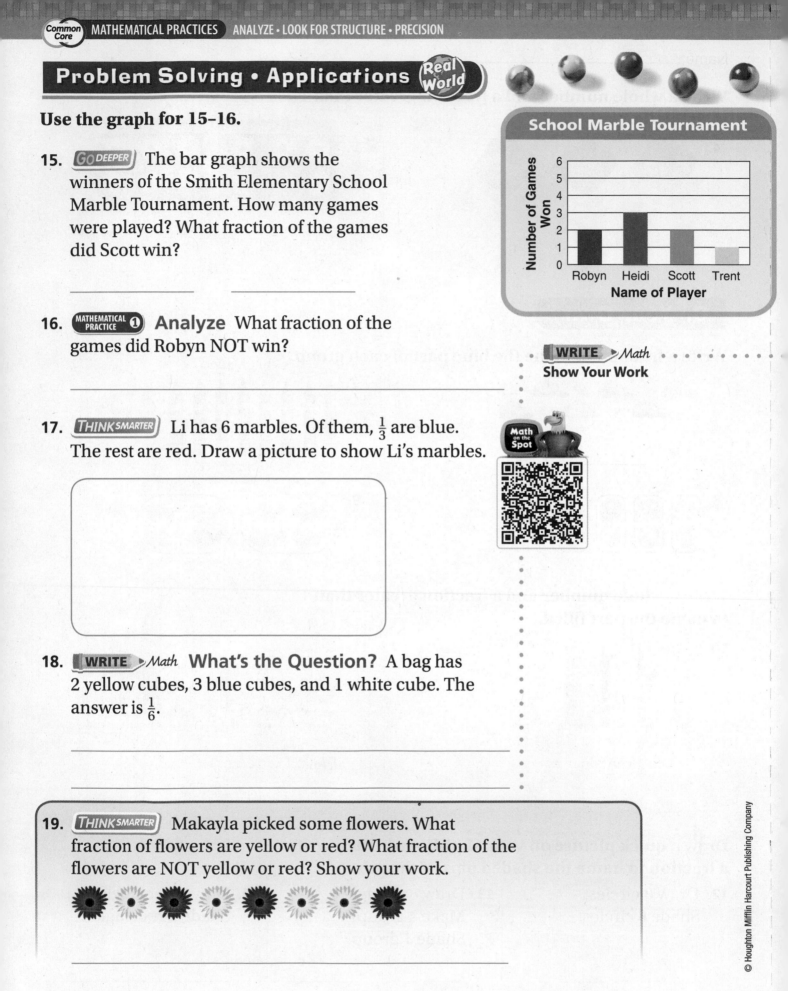

15. **GO DEEPER** The bar graph shows the winners of the Smith Elementary School Marble Tournament. How many games were played? What fraction of the games did Scott win?

_____   _____

16. **MATHEMATICAL PRACTICE ①** **Analyze** What fraction of the games did Robyn NOT win?

_____

_____WRITE ▸ Math_____
Show Your Work

17. **THINK SMARTER** Li has 6 marbles. Of them, $\frac{1}{3}$ are blue. The rest are red. Draw a picture to show Li's marbles.

18. **WRITE ▸ Math** **What's the Question?** A bag has 2 yellow cubes, 3 blue cubes, and 1 white cube. The answer is $\frac{1}{6}$.

_____

_____

19. **THINK SMARTER** Makayla picked some flowers. What fraction of flowers are yellow or red? What fraction of the flowers are NOT yellow or red? Show your work.

_____

Fractions of a Group

Write a fraction to name the shaded part of each group.

Common Core

COMMON CORE STANDARD—3.NF.A.1
Develop understanding of fractions as numbers.

1.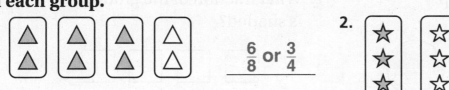

$\dfrac{6}{8}$ or $\dfrac{3}{4}$

2. _____

Write a whole number and a fraction greater than 1 to name the part filled. Think: 1 container = 1

3.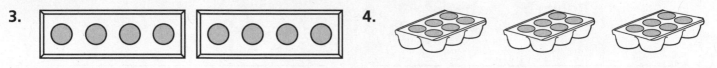

4.

_____ _____   _____

Draw a quick picture. Then, write a fraction to name the shaded part of the group.

5. Draw 4 circles.
   Shade 2 circles.

6. Draw 6 circles.
   Make 3 groups.
   Shade 1 group.

_____   _____

Problem Solving Real World

7. Brian has 3 basketball cards and 5 baseball cards. What fraction of Brian's cards are baseball cards?

8. **WRITE** *Math* Draw a set of objects where you can find a fractional part of the group using the total number of objects and by using subgroups.

_____   _____

Lesson Check (3.NF.A.1)

1. What fraction of the group is shaded?

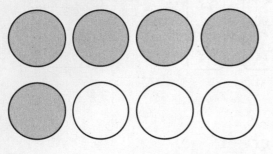

_____

2. What fraction of the group is shaded?

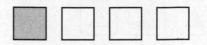

_____

Spiral Review (3.OA.A.3, 3.OA.C.7, 3.NBT.A.2)

3. What multiplication number sentence does the array represent?

_____

_____

_____

4. Juan has 436 baseball cards and 189 football cards. How many more baseball cards than football cards does Juan have?

_____

_____

_____

5. Sydney bought 3 bottles of glitter. Each bottle of glitter cost $6. How much did Sydney spend on the bottles of glitter?

_____

6. Add.

$$\begin{array}{r} 262 \\ +\ 119 \\ \hline \end{array}$$

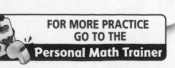
FOR MORE PRACTICE
GO TO THE
Personal Math Trainer

Find Part of a Group Using Unit Fractions

Common Core Number and Operations—
Fractions—3.NF.A.1

Essential Question How can a fraction tell how many are in part of a group?

MATHEMATICAL PRACTICES
MP4, MP5

🔑 Unlock the Problem Real World

Audrey buys a bouquet of 12 flowers. One third of them are red. How many of the flowers are red?

- How many flowers does Audrey buy in all? _____
- What fraction of the flowers are red? _____

🔓 Activity

Materials ▪ two-color counters ▪ MathBoard

- Put 12 counters on your MathBoard.
- Since you want to find $\frac{1}{3}$ of the group, there should be _____ equal groups. Draw the counters below.

- Circle one of the groups to show _____.

  Then count the number of counters in that group.

There are _____ counters in 1 group. $\frac{1}{3}$ of 12 = _____

So, _____ of the flowers are red.

- What if Audrey buys a bouquet of 9 flowers and one third of them are yellow? Use your MathBoard and counters to find how many of the flowers are yellow.

  _____

Math Talk

MATHEMATICAL PRACTICES ❸

Apply How can you use the numerator and denominator in a fraction to find part of a group?

Try This! **Find part of a group.**

Raul picks 20 flowers from his mother's garden.
One fourth of them are purple. How many of the
flowers are purple?

STEP 1 Draw a row of 4 counters.

Think: To find $\frac{1}{4}$, make 4 equal groups.

○   ○   ○   ○

_____

STEP 2 Continue to draw as many rows of
4 counters as you can until you have 20 counters.

_____

STEP 3 Then circle _____ equal groups.

Think: Each group represents $\frac{1}{4}$ of the flowers.

$\frac{1}{4}$ $\frac{1}{4}$ $\frac{1}{4}$ $\frac{1}{4}$

There are _____ counters in 1 group.

$\frac{1}{4}$ of 20 = _____

So, _____ of the flowers are purple.

Share and Show

Math Talk

MATHEMATICAL PRACTICES ⑥

Describe why you count the number of counters in just one of the groups when finding $\frac{1}{2}$ of any number.

1. Use the model to find $\frac{1}{2}$ of 8. _____

   Think: How many counters are in 1 of the 2 equal groups?

Circle equal groups to solve. Count the number of flowers in 1 group.

2. $\frac{1}{4}$ of 8 = _____

3. $\frac{1}{3}$ of 6 = _____

4. $\frac{1}{6}$ of 12 = _____

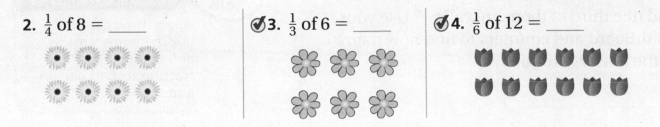

Name _____

On Your Own

Circle equal groups to solve. Count the number of flowers in 1 group.

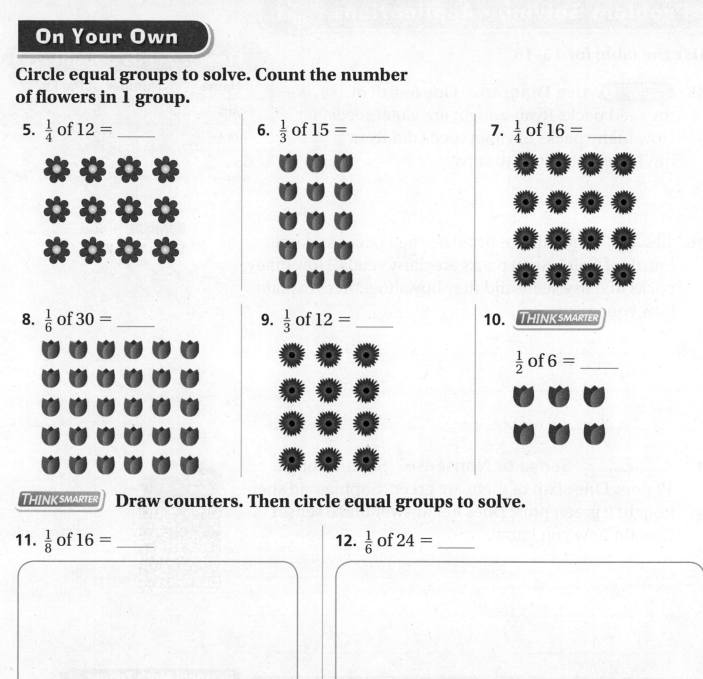

5. $\frac{1}{4}$ of 12 = _____

6. $\frac{1}{3}$ of 15 = _____

7. $\frac{1}{4}$ of 16 = _____

8. $\frac{1}{6}$ of 30 = _____

9. $\frac{1}{3}$ of 12 = _____

10. THINK SMARTER

$\frac{1}{2}$ of 6 = _____

THINK SMARTER **Draw counters. Then circle equal groups to solve.**

11. $\frac{1}{8}$ of 16 = _____

12. $\frac{1}{6}$ of 24 = _____

13. GO DEEPER Gerry has 50 sports trading cards. Of those cards, $\frac{1}{5}$ of them are baseball cards, $\frac{1}{10}$ of them are football cards, and the rest are basketball cards. How many more basketball cards than baseball cards does Gerry have?

14. GO DEEPER Barbara has a mixed garden that has 16 rows of different flowers and vegetables. One-fourth of the rows are lettuce, $\frac{1}{8}$ of the rows are pumpkins, and $\frac{1}{2}$ of the rows are red tulips. The other rows are carrots. How many rows of carrots are in Barbara's garden?

_____

Problem Solving • Applications (Real World)

Use the table for 15–16.

Flower Seeds Bought	
Name	Number of Packs
Ryan	8
Brooke	12
Cole	20

15. **MATHEMATICAL PRACTICE ④ Use Diagrams** One fourth of the seed packs Ryan bought are violet seeds. How many packs of violet seeds did Ryan buy? Draw counters to solve.

_____

16. **GO DEEPER** One third of Brooke's seed packs and one fourth of Cole's seed packs are daisy seeds. How many packs of daisy seeds did they buy altogether? Explain how you know.

_____

_____

_____

WRITE ▸ Math
Show Your Work

17. **THINK SMARTER Sense or Nonsense?** Sophia bought 12 pots. One sixth of them are green. Sophia said she bought 2 green pots. Does her answer make sense? Explain how you know.

_____

_____

_____

18. **THINK SMARTER +** A florist has 24 sunflowers in a container. Mrs. Mason buys $\frac{1}{4}$ of the flowers. Mr. Kim buys $\frac{1}{3}$ of the flowers. How many sunflowers are left? Explain how you solved the problem.

Personal Math Trainer

_____

_____

_____

Find Part of a Group Using Unit Fractions

COMMON CORE STANDARD—3.NF.A.1
Develop understanding of fractions as numbers.

Circle equal groups to solve. Count the number of items in 1 group.

1. $\frac{1}{4}$ of 12 = **3**

○○○○
○○○○
○○○○

2. $\frac{1}{8}$ of 16 = _____

○○○○○○○○
○○○○○○○○

3. $\frac{1}{3}$ of 12 = _____

○○○
○○○
○○○
○○○

4. $\frac{1}{3}$ of 9 = _____

○○○
○○○
○○○

Problem Solving · Real World

5. Marco drew 24 pictures. He drew $\frac{1}{6}$ of them in art class. How many pictures did Marco draw in art class?

_____

6. Caroline has 16 marbles. One eighth of them are blue. How many of Caroline's marbles are blue?

_____

7. **WRITE** ▸*Math* Explain how to find which is greater: $\frac{1}{4}$ of 12 or $\frac{1}{3}$ of 12.

_____

_____

Lesson Check (3.NF.A.1)

1. Ms. Davis made 12 blankets for her grandchildren. One third of the blankets are blue. How many blue blankets did she make?

○ ○ ○
○ ○ ○
○ ○ ○
○ ○ ○

_____

2. Jackson mowed 16 lawns. One fourth of the lawns are on Main Street. How many lawns on Main Street did Jackson mow?

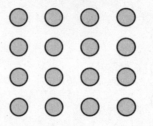

_____

Spiral Review (3.OA.A.7, 3.NBT.A.1, 3.NBT.A.2)

3. Find the difference.

$$509$$
$$-175$$

4. Find the quotient.

$$6\overline{)54}$$

_____

5. There are 226 pets entered in the pet show. What is 226 rounded to the nearest hundred?

_____

6. Ladonne made 36 muffins. She put the same number of muffins on each of 4 plates. How many muffins did she put on each plate?

_____

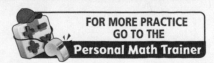

FOR MORE PRACTICE GO TO THE Personal Math Trainer

Problem Solving • Find the Whole Group Using Unit Fractions

Essential Question How can you use the strategy *draw a diagram* to solve fraction problems?

Common Core   **Number and Operations—Fractions—3.NF.A.1**
MATHEMATICAL PRACTICES
MP1, MP4, MP5, MP6

🔑 Unlock the Problem Real World

Cameron has 4 clown fish in his fish tank. One third of the fish in the tank are clown fish. How many fish does Cameron have in his tank?

Use the graphic organizer to help you solve the problem.

Read the Problem	**Solve the Problem**
What do I need to find? I need to find _____ are in Cameron's fish tank.	**Describe how to draw a diagram to solve.** The denominator in $\frac{1}{3}$ tells you that there are _____ equal parts in the whole group. Draw 3 circles to show _____ equal parts. Since 4 fish are $\frac{1}{3}$ of the whole group,
What information do I need to use? Cameron has _____ clown fish. _____ of the fish in the tank are clown fish.	draw _____ counters in the first circle. Since there are _____ counters in the first circle, draw _____ counters in each of the remaining circles. Then find the total number of counters.
How will I use the information? I will use the information in the problem to draw a _____ .	 So, Cameron has _____ fish in his tank.

🔓 Try Another Problem

A pet store has 2 gray rabbits. One eighth of the rabbits at the pet store are gray. How many rabbits does the pet store have?

Read the Problem	**Solve the Problem**
What do I need to find?	
What information do I need to use?	
How will I use the information?	

1. **MATHEMATICAL PRACTICE 8** Draw Conclusions How do you know that your answer is reasonable?

_____

_____

2. How did your diagram help you solve the problem? _____

_____

Math Talk

MATHEMATICAL PRACTICES 1

Make Sense of Problems Suppose $\frac{1}{2}$ of the rabbits are gray. How can you find the number of rabbits at the pet store?

Name _____

Unlock the Problem
✓ Circle the question.
✓ Underline important facts.
✓ Put the problem in your own words.
✓ Choose a strategy you know.

1. Lily has 3 dog toys that are red. One fourth of all her dog toys are red. How many dog toys does Lily have?

   First, draw _____ circles to show _____ equal parts.

   Next, draw _____ toys in _____ circle since

   _____ circle represents the number of red toys.

   Last, draw _____ toys in each of the remaining circles. Find the total number of toys.

   So, Lily has _____ dog toys.

2. **THINK SMARTER** What if Lily has 4 toys that are red? How many dog toys would she have?

   _____

3. The pet store sells bags of pet food. There are 4 bags of cat food. One sixth of the bags of food are bags of cat food. How many bags of pet food does the pet store have?

   _____

4. Rachel owns 2 parakeets. One fourth of all her birds are parakeets. How many birds does Rachel own?

   _____

On Your Own

5. **THINK SMARTER**   Before lunchtime, Abigail and Teresa each read some pages from different books. Abigail read 5, or one fifth, of the pages in her book. Teresa read 6, or one sixth, of the pages in her book. Whose book had more pages? How many more pages?

_____

WRITE ▸ *Math* • **Show Your Work**

6. **MATHEMATICAL PRACTICE ②** **Represent a Problem** Six friends share 5 meat pies. Each friend first eats half of a meat pie. How much more meat pie does each friend need to eat to finish all the meat pies and share them equally? Draw a quick picture to solve.

_____

7. **GO DEEPER**   Braden bought 4 packs of dog treats. He gave 4 treats to his neighbor's dog. Now Braden has 24 treats left for his dog. How many dog treats were in each pack? Explain how you know.

_____

_____

_____

_____

8. **THINK SMARTER**   Two hats are $\frac{1}{3}$ of the group. How many hats are in the whole group?

_____ hats

Problem Solving • Find the Whole Group Using Unit Fractions

COMMON CORE STANDARD—3.NF.A.1
Develop understanding of fractions as numbers.

Draw a quick picture to solve.

1. Katrina has 2 blue ribbons for her hair.
   One fourth of all her ribbons are blue. How
   many ribbons does Katrina have in all?

   _____ **8 ribbons** _____

2. One eighth of Tony's books are mystery
   books. He has 3 mystery books. How
   many books does Tony have in all?

   _____

3. Brianna has 4 pink bracelets. One third
   of all her bracelets are pink. How many
   bracelets does Brianna have?

   _____

4. Ramal filled 3 pages in a stamp album.
   This is one sixth of the pages in the album.
   How many pages are there in Ramal's stamp album?

   _____

5. Jeff helped repair one half of the bicycles in a bike
   shop last week. If Jeff worked on 5 bicycles, how many
   bicycles did the shop repair last week?

   _____

6. **WRITE** ▸*Math* Write a problem about a group of objects in
   your classroom. Tell how many are in one equal part of the
   group. Solve your problem. Draw a diagram to help you.

   _____

   _____

Lesson Check (3.NF.A.1)

1. A zoo has 2 male lions. One sixth of the lions are male lions. How many lions are there at the zoo?

2. Max has 5 red model cars. One third of his model cars are red. How many model cars does Max have?

Spiral Review (3.OA.A.3, 3.NBT.A.1, 3.NBT.A.2, 3.NF.A.1)

3. There are 382 trees in the local park. What is the number of trees rounded to the nearest hundred?

4. The Jones family is driving 458 miles on their vacation. So far, they have driven 267 miles. How many miles do they have left to drive?

$$458$$
$$- 267$$

5. Ken has 6 different colors of marbles. He has 9 marbles of each color. How many marbles does Ken have in all?

6. Eight friends share two pizzas equally. How much of a pizza does each friend get?

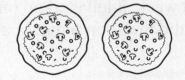

FOR MORE PRACTICE
GO TO THE
Personal Math Trainer

Name _____

✔ Chapter 8 Review/Test

1. Each shape is divided into equal parts. Select the shapes that show thirds. Mark all that apply.

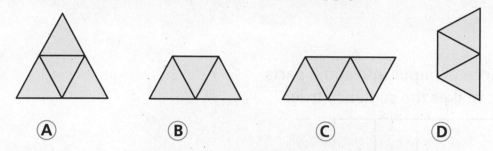

Ⓐ   Ⓑ   Ⓒ   Ⓓ

2. What fraction names the shaded part of the shape?

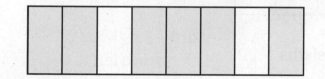

- Ⓐ   8 sixths
- Ⓑ   8 eighths
- Ⓒ   6 eighths
- Ⓓ   2 sixths

3. Omar shaded a model to show the part of the lawn that he finished mowing. What fraction names the shaded part? Explain how you know how to write the fraction.

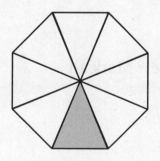

_____

_____

 Assessment Options
Chapter Test

4. What fraction names point *A* on the number line?

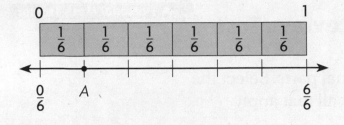

_____

5. Jamal folded this piece of paper into equal parts.
Circle the word that makes the sentence true.

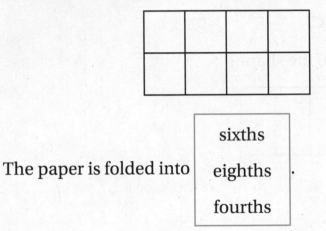

The paper is folded into | sixths
eighths .
fourths

6. Caleb took 18 photos at the zoo. One sixth of his
photos are of giraffes. How many of Caleb's photos are
of giraffes?

_____ photos

7. Three teachers share 2 packs of paper equally.

How much paper does each teacher get? Mark all
that apply.

Ⓐ 3 halves of a pack

Ⓑ 2 thirds of a pack

Ⓒ 3 sixths of a pack

Ⓓ 1 half of a pack

Ⓔ 1 third of a pack

8. Lilly shaded this design.

Select one number from each column to show
the part of the design that Lilly shaded.

Numerator	Denominator
○ 1	○ 3
○ 3	○ 4
○ 5	○ 5
○ 6	○ 6

9. Marcus baked a loaf of banana bread for a party.
He cut the loaf into equal size pieces. At the end of the
party, there were 6 pieces left. Explain how you can find
the number of pieces in the whole loaf if Marcus told you
that $\frac{1}{3}$ of the loaf was left. Use a drawing to show your
work.

_____

_____

_____

_____

10. The model shows one whole. What fraction of the model is NOT shaded?

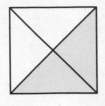

_____

11. Together, Amy and Thea make up $\frac{1}{4}$ of the midfielders on the soccer team. How many midfielders are on the team? Show your work.

_____ midfielders

12. Six friends share 4 apples equally. How much apple does each friend get?

_____

13. Each shape is 1 whole.

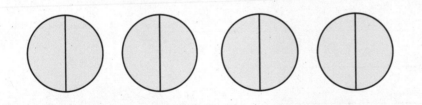

For numbers 13a–13e, choose Yes or No to show whether the number names the parts that are shaded.

13a. 4   ○ Yes   ○ No

13b. 8   ○ Yes   ○ No

13c. $\frac{8}{2}$ ○ Yes   ○ No

13d. $\frac{8}{4}$ ○ Yes   ○ No

13e. $\frac{2}{8}$ ○ Yes   ○ No

14. Alex has 3 baseballs. He brings 2 baseballs to school. What fraction of his baseballs does Alex bring to school?

_____

15. GO DEEPER Janeen and Nicole each made fruit salad for a school event.

Part A

Janeen used 16 pieces of fruit to make her salad. If $\frac{1}{4}$ of the fruits were peaches, how many peaches did she use? Make a drawing to show your work.

_____ peaches

Part B

Nicole used 24 pieces of fruit. If $\frac{1}{6}$ of them were peaches, how many peaches in all did Janeen and Nicole use to make their fruit salads? Explain how you found your answer.

_____

_____

16. There are 8 rows of chairs in the auditorium. Three of the rows are empty. What fraction of the rows are empty?

_____

17. Tara ran 3 laps around her neighborhood for a total of 1 mile yesterday. Today she wants to run $\frac{2}{3}$ of a mile. How many laps will she need to run around her neighborhood?

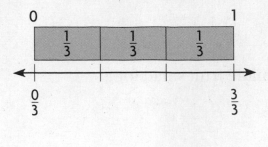

_____ laps

18. Gary painted some shapes.

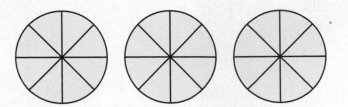

Select one number from each column to show a fraction greater than 1 that names the parts Gary painted.

Numerator	Denominator
○ 3	○ 3
○ 4	○ 4
○ 8	○ 8
○ 24	○ 24

Personal Math Trainer

19. **THINK SMARTER +** Angelo rode his bike around a bike trail that was $\frac{1}{4}$ of a mile long. He rode his bike around the trail 8 times. Angelo says he rode a total of $\frac{8}{4}$ miles. Teresa says he is wrong and that he actually rode 2 miles. Who is correct? Use words and drawings to explain how you know.

_____

_____

_____

_____

_____